Biotechnology, Plant Breeding, Genetics and Biostatistics

About the Book

Plant breeding started with sedentary agriculture and particularly the domestication of the first agricultural plants, a practice which is estimated to date back 9,000 to 11,000 years. Initially early farmers simply selected food plants with particular desirable characteristics, and employed these as progenitors for subsequent generations, resulting in an accumulation of valuable traits over time.

Gregor Mendel's experiments with plant hybridization led to his establishing laws of inheritance. Once this work became well known, it formed the basis of the new science of genetics, which stimulated research by many plant scientists dedicated to improving crop production through plant breeding.

About the Author

Saba Hussain, is post Graduate in Biotechnology from Dr.Bhimrao Ambedkar University, Agra, Uttar Pradesh and Submitted PhD in Biotechnology from C.M.J.University Shillong, Assam. Presently she is working as a Assitant Professor in Biotechnology Department at Patna Science College of Pharmaceutical Education & Research, Patna, Bihar. She has attented various seminars and Conferences and Presented International papers.

Biotechnology, Plant Breeding, Genetics and Biostatistics

Saba Hussain

AMIGA PRESS INC.
DELHI (INDIA)

AMIGA PRESS INC.
H.O.: 28-A, Indra Park,
Chander Nagar, Delhi-110051 (India)
Tel.: 011-22504028, +91-98736-57000
E-mail: amigapress@gmail.com
amigaeditorial@gmail.com
amigapub@gmail.com
info@amigapress.co.in
Website : www.amigapress.co.in

This edition is published by Amiga Press Inc.

ISBN: 978-93-87295-31-5

First Published 2020

Printed and bound in India by Replika Press Pvt. Ltd.

Preface

Plant breeding is the art and science of changing the traits of plants in order to produce desired characteristics. Plant breeding can be accomplished through many different techniques ranging from simply selecting plants with desirable characteristics for propagation, to more complex molecular techniques. Plant breeding has been practiced for thousands of years, since near the beginning of human civilization.

It is now practiced worldwide by individuals such as gardeners and farmers, or by professional plant breeders employed by organizations such as government institutions, universities, crop-specific industry associations or research centers. International development agencies believe that breeding new crops is important for ensuring food security by developing new varieties that are higher-yielding, resistant to pests and diseases, drought-resistant or regionally adapted to different environments and growing conditions.

Plant breeding started with sedentary agriculture and particularly the domestication of the first agricultural plants, a practice which is estimated to date back 9,000 to 11,000 years. Initially early farmers simply selected food plants with particular desirable characteristics, and employed these as progenitors for subsequent generations, resulting in an accumulation of valuable traits over time.

Gregor Mendel's experiments with plant hybridization led to his establishing laws of inheritance. Once this work became well known, it formed the basis of the new science of genetics, which stimulated research by many plant scientists dedicated to improving crop production through plant breeding.

Modern plant breeding is applied genetics, but its scientific basis is broader, covering molecular biology, cytology, systematics, physiology, pathology, entomology, chemistry, and statistics. It has also developed its own technology.

Modern plant breeding may use techniques of molecular biology to select or, or in the case of genetic modification, to insert, desirable traits into plants. Application of biotechnology or molecular biology is also known as molecular breeding.

– Author

Contents

1

Plant Tissue Culture Techniques

Fungi and bacteria grow on the plant surface and contaminate the culture media when they are not adequately eliminated. The process of explant introduction into in vitro conditions depends mainly on the disinfecting phase. Problems in handling produce contamination of the in vitro plant and so, a similar disinfection process to that of the in vitro introduction technique is recommended.

However, it is possible to use antibiotics included in the medium, but just temporarily, while the plantlets are growing. Among the bacterial contaminants are Bacillus sp., Erwin/a sp., Pseudomonas sp., etc. In the case of systemic infections, the use of disinfectants is not effective, because the pathogens are located in the vascular system. In this case, it is recommended to use specific antibiotics and meristems or bud cuttings.

On the other hand, the process of in vitro introduction involves the use of explants in different physiological stages, as dormant buds, which require growth regulators for example, Gibberellic acid) to stimulate and accelerate bud growth. During the process of in vitro introduction there is a close connection between the greenhouse and the laboratory; that is why the necessary precautions should be taken to avoid the entrance of contaminants into the laboratory.

- In the greenhouse
  - Take cuttings from the mother plant, or buds from greenhouse tuber sprouts.
- In the laboratory
  - Immerse the cuttings for 10 minutes in a beaker conveniently labeled with the accession name, and containing a solution of 0.5 per cent

acaricide with three drops/I of Tween 20.

- Throw away the acaricide solution and rinse the nodes with tap water
- Prepare the laminar flow chamber.
- Immerse the cuttings in 700/o alcohol for 30 seconds and take the beakers immediately to the laminar flow chamber

• In the laminar flow chamber
  - Eliminate the alcohol and replace it with a 2.50/o solution of calcium hypochlorite. Keep the cuttings immersed during 15 minutes.
  - Throw away the hypochlorite, and rinse the cuttings three times with sterile water
  - Keep the nodes immersed in sterile water until bud extraction.
  - Dissect the buds on a sterile plate and place them in the culture medium.

MICROPROPAGATION

In vitro plantlets, which are free of pathogens, are used as initial material for potato and sweetpotato seed programmes. The methods used in these micropropagation programmes mainly depend on their production volume and the available infrastructure.

In the case of potato micropropagation, the basic methods have been already described. They have been verified in many institutions and they are based on the rapid growth of individual node cuttings, or stems with multiple node cuttings. Afterwards, the basic micropropagation methods are described.

NODE MICROPROPAGATION

This method is based on the principle that the node of an in vitro plantlet place dinan appropriate culture medium will induce the development of the axillary bud, resulting in a new in vitro plantlet. This type of propagation promotes the development of a pre-existent morphological structure. The nutritional and hormonal condition of the medium breaks the dormancy of the axillary bud and promotes its rapid development.

Callus formation and plant regeneration must be avoided because they tend to affect the genetic stability of the genotype. Under room-controlled conditions micropropagation is fast. Each node planted in a propagation medium will produce a plantlet which will occupy the full length of the test tube, after approximately four weeks for potato, and six weeks for sweet potato. The resultant in vitro plantlets may be transplanted to in vitro conditions in small pots in the greenhouse.

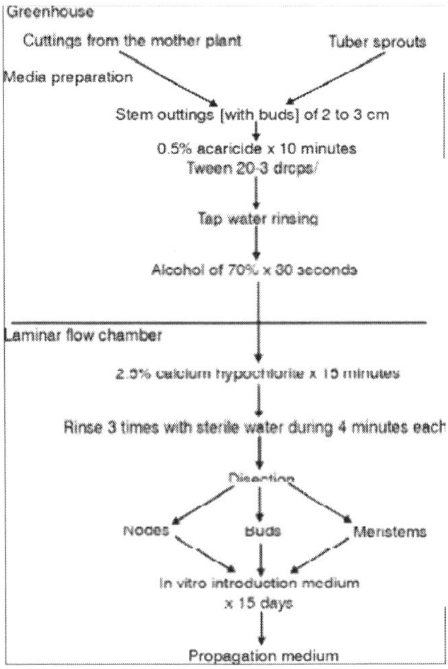

Fig. Procedure for in vitro Introducction

MICROPROPAGATION BY NODE CUTTINGS IN A LIQUID MEDIUM

This technique is applied both with potato and sweetpotato to produce a large number of nodes rapidly. Stem cuttings with 5 to 8 nodes are prepared by removing both the apex and the root of the in vitro plant to be propagated. The stems are placed in the corresponding propagation liquid medium. It is also possible to use isolated nodes: the nodes will sprout and new plantlets will develop over a period of 3 to 4 weeks.

Micropropagation Procedure

- Sterilize petri dishes (placed in paper bags or comets) and prepare the laminar flow chamber by disinfecting the internal surfaces with alcohol.

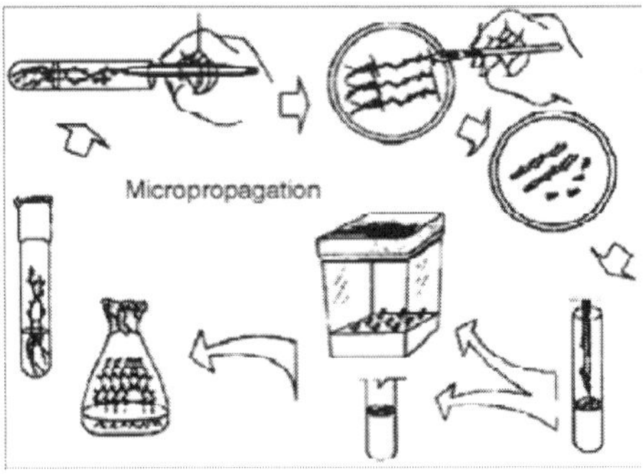

Fig. Potato Micropropagation Process Scheme

Sterilize the tools with an instrument sterilizer and place them on a sterile dish.

- Open the tube, take off the plantlet and place it on a petri dish with the help of forceps.
- Remove the leaves and cut the nodes.
- Open a tube containing fresh sterile medium and place a node inside, trying to plunge it slightly into the medium with the bud up. Close the tube.
- Seal the tube with a gas-permeable plastic tape (parafilm or saram wrap) and label it correctly. It is recommended to place two explants in 16 × 125 mm tubes, three in 18 × 150 mm tubes, five in 25 × 150 mm tubes, and 20-30 in magenta vessels.

COMMON PROBLEMS IN MICROPROPAGATION

Some problems may appear in tissue culture according to the crop or variety. To solve them, it is necessary to apply one or several preventing/ solving methods such as:

Phenolization

The explants frequently become brown or blackish shortly after isolation. When this occurs, growth is inhibited and the tissue generally dies. The young tissues are less susceptible to darkening than the more mature ones.

Prevention

The tissue darkening– mainly that of the recently isolated explants and that of the medium–may be generally prevented by:

- Removing the phenolic compounds produced by dispersion. A b s orption by m e an s of a c tiva te d c a rbon . A bsorpt ion by polyvinilpyrolidone (PVP).
- Modifying the redox potential. Reducing agents: ascorbic acid, citric acid, L-cisteine HCL, ditriotreitol, glutation and mercaptoethanol Less availability of oxygen: stationary liquid media.
- Inactivating the phenolase enzymes. Chelating agents: NaFeEDTA, EDTA, diethyldithiocarbamate, dimethyl-dithiocarbamate
- Reducing the phenolasic activity and the availability of substrate. Low pH Darkness

Absence of Rooting

The explants can naturally form roots during propagation, without an additional rooting stage, as with the potato. However, some wild potato species may show root production deficiency. Rooting may be induced by incorporating auxins, such as IAA, NAA, and IBA, or activated carbon to the culture medium.

POTATO IN VITRO TUBERIZATION

Most of the potato microtubers are used in seed programmes in Europe, where large amounts (hundred of thousands) of pre-basic seed of a few varietie s are produc ed. By mea ns of this technique, microtubers are produced and stored, and it is possible to store thousands of them in a small area (in humid containers at 4oC) for long periods of time. The microtuber induction is produced through a stress effect by the CCC (chlorocholine chloride), BAP (B-benzylaminopurine) and Sucrose, which under darkness will produce from 3 to 4 microtubers per plant, according to the variety.

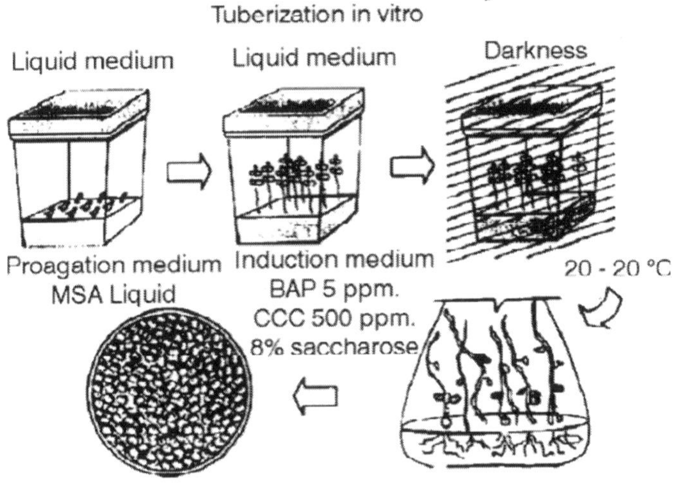

Fig. Poato in Vitro Tuberization Process Scheme

In the beginning, micro tubers were used as an alternative for germplasm distribution and in vitro conservation. Curing the plantlet in vitro multiplication process, usually there is a multiplication of higher amounts to those required in the green house and so, after the transference, some magentas are left over. These plantlets may be used for microtuber induction.

In accordance with the indicated procedures, the induction medium is added to the magentas and then they are placed in the dark room. After three months the microtubers will be produced. These may be harvested and transferred to sterile containers (at 4°C), where they can be maintained up to 10 months, or used in the next campaign instead of in vitro plantlets.

These microtubers can also be used as a reserve, in case in vitro material is contaminated or dies, because of temperature or handling effects. The high number and bigger size of the microtubers increases the success rates for the transfer to the greenhouse.

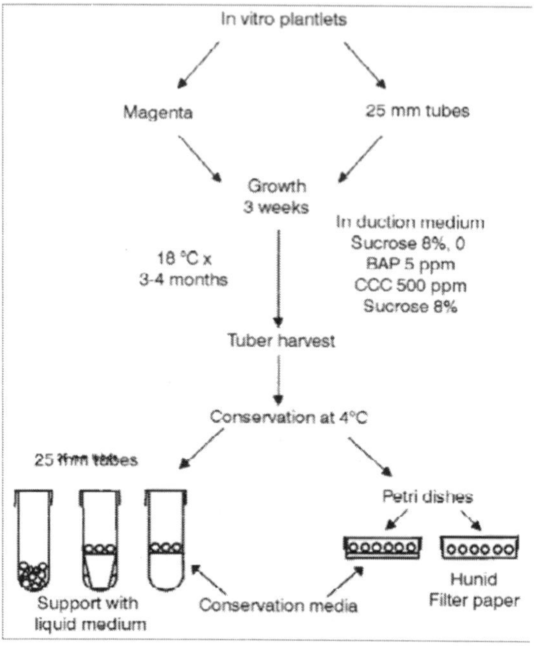

Fig. Potato in Vitro Tuberization Process Scheme

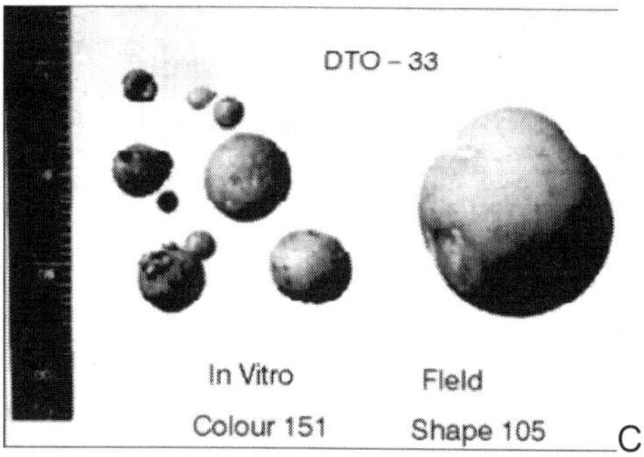

Fig. Comparison, in Size and Shape of in Vitro Microtubers with a Normal one from the Field

MEDIUM TERM GERMPLASM MAINTENANCE

The in vitro germplasm maintenance under normal growth conditions requires a series of transfers of the plantlets into a fresh medium. This leads to a consumption of time, increases the possibility of losses because of material contamination during successive sub-cultures, causes a loss of material by human error or failure of some equipment, and demands more labour A way to avoid these problems is through limitation, restriction or inhibition of growth. This approach consists of growth speed reduction by modifying the physical or chemical conditions of the culture, and it is effective for a short or medium term period.

In Vitro Conservation Methods

These consist in maintaining the cultures (buds, plantlets derived from nodes or directly from meristems) under physical (environmental factors] or chemical (culture medium composition) stressed conditions that make it possible to extend, as much as possible, the interval of transference into the fresh media, without affecting the viability of the cultures. The methods to reduc e the in vitr o cultiva te d plant growth include the reduction of temperature and light during storage, the incorporation of growth retardants in the medium, and the induction of osmotic stress in the medium, or a combination of all these.

Temperature

Temperature reduction has been the most used way to curb culture growth. Most of the in vitro cultures are maintained at temperatures between 12 and 20°C: at lower temperatures, the growth rate decreases but the reduction depends on the species.

Nutrients Concentration

The reduction of the carbohydrate concentration and the nitrogenous components of the nutritive medium may affect the growth rates. In addition, the continuous absorption of these nutrients during plantet growth will bring a nutritional defficiency, which could produce a premature death of the plants.

Use of Growth Regulators

The abscisic acid (ABA), phosphon-D, maleic hydrazide, and succinic acid are some of the most frequently used growth regulators.

Osmotic Concentration

Growth limita tion caused by osmotic concentration is due to the reduction of the water and nutrients absorbed from the medium. For example, at high concentrations, sucrose acts osmotically and it is highly metabolized. No metabolized osmotics, such as Manitol and Sorbitol, are possibly more efective than sucrose in culture growth limitation.

Evaluation of in Vitro Preserved Material

To evaluate material under these conditions, some important facts about in vitro maintenance such as genetic viability and stability should be considered. The viability evaluation of the in vitro cultures must be systematic. In slow growth conditions, when the sub-culture or transfer period extends during months or years, the frequency of the culture evaluation increases. The most important characteristics to be evaluated in the low-growth cultivars storage of apical buds are: contamination, leaf senescence, the number of green sprouts, the number of viable nodes in relation to the stem length, the presence or not of roots, and callus formation.

MAINTENANCE OF ACCESSIONS IN SEED PROGRAMME

In a seed programme, a group of accessions free of virus is maintained for pre-basic seed production: however, most of them are not propagated in the greenhouse so they are maintained in vitro to be used in the future. The continuous propagation of the in vitro plantlets damages the material, mainly if we consider that environmental conditions are not adequate. The alternative is to maintain the accessions in conservation media. Each accession must be maintained in test tubes with five replications to avoid possible losses. The maintenance conditions have been tested in CIP's potato collection that consists of more than 5,000 accessions, by means of which it is possible to assure the normal recovery of the material after using stress producers.

The plantlets used in each campaign must come from the maintenance phase. Then, they will be propagated in normal media where their growth will be reestablished, and the elected procedures for plantlet multiplication in the green house will continue. Each season must be initiated with this material to start off with strong plants. In addition, the plantlets taken for multiplication must be replaced, trying to maintain 5 tubes per accession in the conservation media.

The sub-cultures in the conservation medium are carried out approximately every one or two years. The conservation medium renovation must pass through a previous sub-culture in the propagation media to rejuvenate the explants.

VIRUS ERADICATION THROUGH MERISTEMS CULTURE AND THERMOTHERAPY

If a healthy plant is sown in the field, it is exposed to infections caused by pathogens as nematodes, fungi, bacteria, phytoplasms, virus and viroids, which have a negative effect on yield, and in some cases may kill the plants. However, not all the plant cells may become infected. A group of cells, which are in a continuous non-differentiated multiplication, are virus-free : the meristem. The in vitro meristems culture, together with growth at high temperatures produces potato plantlets free from viruses in more than 90 per cent of planted meristems. This routine method was established in the International Potato Centre to obtain virus-free plantlets for national and international distribution. The in vitro maintenance of virus-free plants provides the possibility to maintain all the time a bank of healthy and more vigourous plants, and with a more accelerated growth, than the infected ones. In a seed programme, it is essential to initiate this work with virus-free plantlets since, this will affect the seed quality and the yield as well. The virus cleaning procedure is too long and expensive: that is why CIP, through its germ plasm distribution programme, has provided a list of the principal virus-free potato varieties to all users.

PROCEDURE

- Approximately 18 to 20 plantlets (virus-infected) are propagated in magentas.
- After a growth period of 20 to 25 days, or when the plantlets are 4-5 cm high, they are placed in the thermotherapy chamber. The growth conditions are: 16 hours of light 34°C; 8 hours of darkness 32°C
- The magentas are maintained in the thermotherapy chamber for one month.
- Afterwards, the magentas are taken out of the chamber, and the outside is cleaned with 980° alcohol. Then they are introduced to the culture room.
- Meristems are obtained as follows:
  - Cut the apical portion and remove the leaves that cover the meristem (approximately 3 to 4 leaves); the meristem is observed with a prominent leaf primordium. Remove the meristem with part of the leaf primordium; cut only the translucent portion. Use a new knife.

Place the menistem in the culture medium. Be sure that the meristem is in the tube.

- Evaluate meristem growth and transfer them to fresh media if necessary.
- Each meristem that originates a plant is called <‘line», which will be labeled according to the accession it belongs to. For example, Yellow Line 1, Yellow Line 2, Yellow Line 3, etc.
- Five tubes with several plants are propagated to evaluate the potato virus PSTVd, PVT, to determine the host range, the morphologic evaluation, and the in vitro maintenance.
- The results of each evaluation are obtained, and the infected material is replaced with clean material.

Temperature

The growth of the plantlets at temperatures higher than 20°C is accelerated (Do not use temperatures higher than 30°C).

CULTURED CELLS AND TISSUES

The large scale commercial propagation of plant material based on plant tissue culture was pioneered in the USA. During the last thirty years, tissue culture-based plant propagation has emerged as one of the leading global agro-technologies. Between 1986 and 1993, the worldwide production of tissue cultured plants increased 50 per cent. In 1993, the production was 663 million plants. By 1997, production had risen to 800 million plants. During 1990–1994, the micropropagation industry declined in Europe, mainly due to production shifting to developing countries, but since, then because of the demand for high quality and number, production in European countries has increased.

Since, 1995, production has increased by 14 per cent in Asian countries, mainly due to the market entry of China, while the increase in South and Central America wa s from production in Cuba. More recently, some companies from Israel, the USA and UK have shifted their production requirements to Costa Rica and India.

Tissue cultured plants have as yet to reach many growers and farmers in the developing countries. The primary advantage of micropropagation is the rapid production of high quality, disease-free and uniform planting material. The plants can be multiplied under a controlled environment, anywhere, irrespective of the season and weather, on a year-round basis. Production of high quality and healthy planting material of ornamentals, and forest and fruit trees, propagated from vegetative parts, has created new opportunities in global trading for producers, farmers, and nursery owners, and for rural employment. Micropropagation technology is more expensive than the conventional methods of plant propagation, and requires several types of skills. It is a capital-intensive industry, and in some cases the unit cost per plant becomes unaffordable. The major reasons are cost of production and know-how. During the early years of the technology, there were difficulties in selling tissue culture products because the conventional planting material was much cheaper.

Now this problem has been addressed by inventing reliable and cost effective tissue culture methods without compromising on quality. This requires a constant monitoring of the input costs of chemicals, media, energy, labour and capital. In the industrialized countries, labour is the main factor that contributes to the high cost of production of tissue-cultured plants.

To reduce such costs, some steps can be partially mechanized, e.g., use of a peristaltic pump for medium dispensing, and of dishwashers for cleaning containers. In the less developed countries of Africa, Asia, and Latin America, where labour is relatively cheaper, consumables such as media, culture containers, and electricity make a comparatively greater contribution to production costs.

For example, the cost of medium preparation (chemicals, energy and labour) can a ccount for 30–35 per cent of the micropropagated plant production. However, automated production processes based on pre-sterilized membrane capsules, bioreactors, mechanized explant transfer, and container sealing are not commercially viable propositions in many developing countries. Therefore, low cost alternatives are needed to reduce production cost of tissue-cultured plants.

Many of the low cost technology options described in this publication ca n be incorporated in various steps of plant micropropagation. The occasional tissue culture generated variants (somaclones) and rare spontaneous bud mutants as well as those obtained from induced mutations can also be propagated by deployment of low cost techniques described. Micropropagation, in combination with radiation-induced mutations, speeds up the recovery, multiplication and release of improved varieties in vegetatively propagated plants. Hence, low cost technology will be of great value for large scale plant multiplication of mutants of many fruits, shrubs, flowers and forest trees that are conventionally vegetatively propagated.

PLANT TISSUE CULTURE

Plant tissue culture refers to growing and multiplication of cells, tissues and organs of plants on defined solid or liquid media under aseptic and controlled environment. The commercial technology is primarily based on micropropagation, in which rapid proliferation is achieved from tiny stem cuttings, axillary buds, and to a limited extent from somatic embryos, cell clumps in suspension cultures and bioreactors. The cultured cells and tissue can take several pathways. The pathways that lead to the production of true-to-type plants in large numbers are the preferred ones for commercial multiplication. The process of micropropagation is usually divided into several stages i.e., prepropagation, initiation of explants, subculture of explants for proliferation, shootin g a nd rootin g, and hardening. These stages a re universally applicable in large-scale multiplication of plants. The delivery of hardened small micropropagated plants to growers and market also requires extra care.

Plant tissue culture refers to growing and multiplication of cells, tissues and organs on defined solid or liquid media under aseptic and controlled environment. Plant tissue culture technology is being widely used for large- scale plant multiplication. The commercial technology is primarily based on micropropagation, in which rapid proliferation is achieved from tiny stem cuttings, axillary buds, and to a limited extent from somatic embryos, cell clumps in suspension cultures and bioreactors.

EXPLANT SOURCE

Plant tissue cultures are initiated from tiny pieces, called explants, taken from any part of a plant. Practically all parts of a plant have been used successfully as a source of explants. In practice, the "explant" is removed surgically, surface sterilized and placed on a nutrient medium to initiate the moth er culture, that is multiplied repeatedly by subculture . The following plant parts are extensively used in commercial micropropagation. Shoot-tip and meristem-tip culture: Shoots develop from a small group of cells known as shoot apical meristem. The apical meristem maintains itself, gives rise to new tissues and organs, and communicates signals to the rest of the plant. Shoot-tips and meristem-tips are perhaps the most popular source of expla nts to initiate tissue cultures. The shoot a pex ex plant measures between 100 to 500 μ m and includes the apical meristem with 1 to 3 leaf primordia. The apical meristem of a shoot is the portion lying distal to the youngest leaf primordium, and is ca.100 μ??m in diameter and 250μ?m in length with 800-1200 cells. In practice, shoot-tip explants between 100 to 1000 μ m are cultured to free plants from viruses. Even explants larger than 1000 μ m have been frequently used. The term "meristem-tip culture" has been suggested to distinguish the large explants from those used in conventional propagation.

Nodal or axillary bud culture: This consists of a piece of stem with axillary bud culture with or without a portion of shoot. When only the axillary bud is taken, it is designated as "axillary bud" culture. Floral meristem and bud culture: Such explants are not commonly used in commercial propagation, but floral meristems and buds can generate complete plants. Other sources of explants: In some plants, leaf discs, intercalary meristems from nodes, small pieces of stems, immature zygotic embryos and nucellus have also been used as explants to initiate cultures.

Cell suspension and callus cultures: Plant parts such as leaf discs, intercalary meristems,-stem-pieces, immature embryos, anthers, pollen, microspores and ovules have been cultured to initiate callus. A callus is a mass of unorganized cells, which in many cases, upon transfer to suitable medium, is capable of giving rise to shoot-buds and somatic embryos, which then form complete plants. Such calli on culture in liquid media on shakers are used for initiating cell suspensions. Liquid suspension cultures maintained on mechanical shakers achieve fast and excellent multiplication rates. However, in commercial micropropagation, calli are cultured mostly in bottles and flasks kept on semi-solid or liquid media. To a limited extent, bioreactors have become popular for somatic embryogenic cultures. It is considered that some day robotics could be adapted to bioreactorbased micropropagation.

PATHWAYS OF CULTURED CELLS AND TISSUES

The cultured cells and tissue can take several pathways to produce a complete plant. Among these, the pathways that lead to the production of true-to-type plants in large numbers are the popular and preferred ones for commercial multiplication. The following terms have been used to describe various pathways of cells and tissue in culture.

Regeneration and Organogenesis

In this pathway, groups of cells of the apical meristem in the shoot apex, axillary buds, root tips, and floral buds are stimulated to differentiate and grow into shoots and ultimately into complete plants. In many cases, the axillary buds formed in the culture undergo repetitive proliferation, and produce large number of tiny plants.

The plants are then separated from each other and rooted either in the next stages of micropropagation or in vivo (in trays, small pots or beds in glasshouse or plastic tunnel under relatively high humidity). The explants culture do relatively high amounts of auxin (e.g., (2,4 - D , 2 , 4- dichlorophenoxyacetic acid) form an unorganized mass of cells, called callus. The callus can be further sub-cultured and multiplied.

The callus shaken in a liquid medium produces cell suspension, which can be subcultured and multiplied into more liquid cultures. The cell suspensions form cell clumps, which eventually form calli and give rise to plants through organogenesis or somatic embryogenesis. In some cases, explants e.g., leaf-discs and epidermal tissue can also generate plants by direct organogenesis and somatic embryogenesis without intervening callus formation, e.g., in orchardgrass. Dactylis glomerata L.. In organogenesis the cultured plant cells and cell clumps (callus) and mature differentiated cells (microspores, ovules) and tissues (leaf discs, inter-nodal segments) are induced to differentiate into complete plants to form shoot buds and eventually shoots, and rooted to form complete plants.

Somatic Embryogenesis

In this pathway, cells or callus cultures on solid media or in suspension cultures form embryo-like structures called somatic embryos, which on germination produce complete plants. The primary somatic embryos are also c a pable of p r od ucin g more em br yo s through se c o ndary s o ma ti c embryogenesis. Although, somatic embryogenesis has been demonstrated in a very large number of plants and trees, the use of somatic embryos in large-scale commercial production has been restricted to only a few plants, such as carrot, date palm, and a few forest trees.

Somatic embryos are produced as adventitious structures directly on explants of zygotic embryos, from callus and suspension cultures. Somatic embryos and synthetic seeds (embryos encapsulated in artificial endosperm) hold potential for large-scale clonal propagation of superior genotypes of heterogeneous plants.

They have also been used in commercial plant production and for the multiplication of parental genotypes in large-scale hybrid seed production. In many species, somatic embryos are morphologically similar to the zygotic em b ryos , alth oug h som e bi ochem ic al , phy s iol ogic a l an d a na tomi cal differences have been documented.

The synthetic auxin, 2,4-D is commonly used for embryo induction. In many angiosperms, e.g., carrot and alfalfa, subculture of cells from 2,4-D containing medium to auxin-free medium is sufficient to induce somatic embryogenesis. The process can be enhanced with the application of osmotic stress, manipulation of medium nutrients, and reducing humidity. Selection of embryogenic cell lines has also been successfully used. For example, selection for unique morphotypes in grapevine cultures allows production of high quality embryos with predicable frequency.

A major problem in large-scale production of somatic embryos is culture synchronization. This is achieved through selecting cells or pre-embryonic cell clusters of certain size, and manipulation of light and temperature, temporary starvation or by adding cell cycle synchronizing chemicals to the medium. Cytokinins seem to play a key role in cell cycle synchronization and embryo induction, proliferation and differentiation. Abscisic acid is crucial in all the stages of somatic development, maturation and hardening.

Synthetic Seeds

The concept of production and utilization of synthetic seeds (somatic embryo as substitutes for true seeds) was first suggested by Murashige in 1977. Synthetic seeds can be produced either as coated or non-coated, desiccated somatic embryos or as embryos encapsulated in hydrated gel (usually calcium alginate).

Successful utilization of synthetic seeds as propagules of choice requires an efficient and reproducible production system and a high percentage of post-planting conversion into vigourous plants. Artificial coats and gel capsules containing nutrients, pesticides and beneficial organisms have long been thought as substitutes for seed coat and endosperm. However, this technology is still in the developmental stage, and currently cannot compete with the other methods of commercial plant propagation.

Process of Micropropagation

The process of plant micro-propagation aims to produce clones (true copies of a plant in large numbers).

The process is usually divided into the following stages:

Stage 0 – Pre-propagation step or selection and pre-treatment of suitable plants.

Stage I – Initiation of explants-surface sterilization, establishment of mother explants.

Stage II – subculture for multiplication/proliferation of explants.

Stage III – shooting and rooting of the explants.

Stage IV – Weaning/hardening.

These stages are universally applicable in large-scale multiplication of plants. The individual plant species, varieties and clones require specific modification of the growth media, weaning and hardening conditions.

A rule of the thumb is to propagate plants under conditions as natural or similar to those in which the plants will be ultimately grown ex-vitro. For example, if a chrysanthemum variety is to be grown under long day-length for flower production, it is better to multiply the material under long-day length at stages III and IV. There is a wide option to undertake production of plant material up to a limited number of stages. For example, many commercial tissue culture companies undertake production up to Stage III, and leave the remaining stages to others.

Pre-Propagation Stage

The pre-propagation sta ge (also called stage 0) requires proper maintenance of the mother plants in the greenhouse under disease-and insect-free conditions with minimal dust. Clean enclosed areas, glasshouses, plastic tunnels, and net-covered tunnels, provide high quality explant source plants with minimal infection. Collection of plant material for clonal propagation should be done after appropriate pretreatment of the mother plants with fungicides and pesticides to minimize contamination in the in vitro cultures. This improves growth and multiplication rates of in vitro cultures.

The control of contamination begins with the pretreatment of the donor plants. They may be prescreened for diseases, isolated and treated to reduce contamination. The explants are then brought to the production facility, surface sterilized and introduced into culture. They may at this stage be treated with antibiotics and fungicides as well as anti-microbial formulations, such as PPM. The explants are then culture indexed for contamination by standard microbiological techniques, which are occasionally supplemented with tests based on molecular biology or other techniques.

Stage I

This stage refers to the inoculation of the explants on sterile medium to initiate aseptic culture. Initiation of explants is the very first step in micropropagation. A good clean explant, once established in an aseptic condition, can be multiplied several times; hence, explant initiation in an aseptic condition should be regarded as a critical step in micropropagation. More than often, explants fail to establish and grow, not due to the lack of a s u it able medium bu t beca us e o f conta m in atio n. The explants ar e transferred to in vitro environment, free from microbial contaminants. The process requires excision of tiny plant pieces and their surface sterilization with chemicals such as sodium hypochlorite, ethyl alcohol and repeated w ash in g with sterile distilled wa te r before and af te r treatment with chemicals. After a short period of culture, usually 3 to 5 days, the contaminated explants are discarded.

The surviving explants showing growth are maintained and used for further subculture. In herbaceous plants e.g., potato, chrysanthemum, carnation, streptocarpus, strawberry, and African violet; the explant sources are meristems, apical- and axillary buds, young seedlings, developing young leaves and petioles, and unopened floral buds. The following low cost options can be adapted to initiate explants:

Sterile Instrument Technique

This method assumes that most of the deep-seated meristems and those covered by leaves or other integuments (e.g., floral bracts) are sterile. In this procedure, the explant is washed with sterile water, rinsed in ethanol, and instruments are sterilized every time they touch the surface of the explant, and the explant is moved to a new location on the dissection stage.

Surface Sterilization Technique

This is by far the most commonly used method. The explants are washed in sterile water, rinsed in ethanol, and surface sterilization is achieved by using chemicals with chlorine base. Calcium or sodium hypochlorite based solutions, 1–3 per cent (v/v) are usually used for soft herbaceous materials. A cheap and ready-made sterilant is 5-7 per cent solution of 'Domestos'- a toilet disinfectant which contains 10.5 per cent v/v sodium hypochlorite, 0.3 per cent sodium carbonate, 10.0 per cent sodium chloride and 0.5 per cent (w/v) sodium hydroxide and a patented thickener). The explants are washed in sterile distilled water before and after sterilization. Other surface sterilants used include mercuric chloride (avoid its use as far as possible, since, it is highly toxic), hydrogen peroxide, and potassium permanganate.

For Soft Tissues

- Wash explants from perennial plants for 1-2 hr in tap water.
  Eliminate this step for material from glasshouse grown plants.
- Wash in sterile distilled water three to four times for 5 to 10 minutes each.
- Dip in 95 per cent ethanol for 3 to 5 seconds.
- Wash once again with sterile distilled water for 5 minutes.
- Surface-sterilize in 5 per cent 'Domestos' (v/v) for 20-25 minutes.
- Wash with sterile distilled water three times for 10 minutes each.
- Drain water droplets by placing on pre-sterilized blotting paper.
- Transfer explants singly to the medium.

N.B.: Sterilize forceps each time to transfer explants to avoid cross-contamination.

For woody stems (e.g., roses, hardy shrubs, and trees):

- Collect stems, shoots, buds and store at 5 oC till needed.
- Rinse in ethanol for 3 to 5 seconds.
- Rinse in 1- per cent sodium hypochlorite (20 per cent bleach) for 10 minutes.
- Place lower parts of stems in flasks in 2 per cent sucrose and 200 PPM 8-hydroxyquinoline citrate at 23+2 oC. For items collected in September/October, add 50-PPM GA3.

After that l0 PPM GA will help break the dormancy:

- Re-cut the bottom of stem and replace the solution after 2 days.
- Excise the softwood from the developed shoots and use material for explants or for rooting.
- Surface-sterilize as in the above protocol.

Do not forget to sterilize forceps and scalpel every time for the transfer of explants to fresh solutions. Use sterile containers in the protocol of surface sterilization. If explants become brown or pale at the end of the protocol, reduce the strength of 'Domestos' to 2.5 per cent. Alternatively, dip explants in 10 per cent 'Domestos' for 2 minutes and then proceed to surface sterilize with 3-5 per cent 'Domestos' for 20 minutes. If basal contamination is observed after 2-3 days of culture, explants can sometimes be rescued by removing the basal end by making a single cut with a sharp scalpel and re-culturing on fresh medium.

Stage II

Stage II is the propagation phase in which the explants are cultured on the appropriate media for multiplication of shoots. The primary goal is to achieve propagation without losing the genetic stability. Repeated culture of axillary and adventitious shoots, cutting with nodes, somatic embryos and other organs from Stage I leads to multiplication of propagules in large numbers. The propagules produced at this stage can be further used for multiplication by their repeated culture. Sometimes it is necessary to subculture the in vitro derived shoots onto different media for elongation.

Stage III

The in vitro shoots obtained at Stage II are rooted to produce complete plants. If the proliferated material consists of bud-like structures (e.g., orchids) or clumps of shoots (banana, pineapple), they should be separated after rooting and not before. Many plants (e.g., banana, pineapple, roses, potato, chrysanthemum, strawberry, mint, several grasses and many more) can be rooted on half-strength-MS medium without any growth-regulators. Good sturdy well-rooted plants are essential for high survival during weaning and later transfer to soil. This stage is labour intensive and expensive. The process of in vitro rooting has been estimated to account for approximately 35–75 per cent of the total cost of production. Efforts should be made to combine rooting and acclimatization stages.

Stage IV

At this stage, the in vitro micropropagated plants are weaned and hardened. This is the final stage of the tissue culture operation after which the micropropagated plantlets are ready for transfer to the greenhouse. Steps are taken to grow individual plantlets capable of carrying out photosynthesis. The hardening of the tissue-cultured plantlets is done gradually from high to low humidity and from low light intensity to high intensity conditions. If grown on solid medium, most of the agar can be removed gently by rinsing with water.

Plants can be left in shade for 3 to 6 days where diffused natural light conditions them to the new environment. The plants are then transferred to an appropriate substrate (sand, peat, compost, etc.), and gradually hardened.

Low-cost options include the use of plastic domes or tunnels, which reduces the natural light intensity and maintains high relative humidity during the hardening process. If the plants are still joined together after rooting, these should be planted as bunches in the soil and separated after 6 to 8 weeks of growth.

DELIVERY TO THE GROWERS

The delivery of the rooted and hardened small micropropagated plants to growers and market requires extra care. In some cases, plant losses can occur during shipment and handling by growers. This is particularly true when the plants are not fully hardened and rooted or not grown for sufficient duration after transfer to soil. Growers should be given clear instructions how to handle the material provided. Apart from the economic loss, poor surviva l of plan ted material erodes the con fidence of growers in the technology. The transfer of individual plants to soil in black plastic or polythene bags is widely used as a low-cost option to provide fully-grown banana plants directly to farmers in many developing countries.

❐

2

Feeding Systems

FIBROUS CROP RESIDUES

FACTORS INFLUENCING NUTRITIVE VALUE

In many developing countries, and in Asia in partic-ular, ruminants are fed on straw from cereal crops (mainly rice and wheat). As population pressure in-creases and the area devoted to food-crop produc-tion is extended the use of crop residues and byprod-ucts for animal feeding will increase. Incountries with specialised livestock production systems, straw is considered of such low feed value that it is often burned, but in developing countries where livestock are integrated with cropping it is a valuable resource.

Draught animals appear to be able to work and to maintain body condition on a diet of mainly straw. Research on the utilisation of straw by ruminants in developed countries has focused on the role of straw as a roughage supplement in a concentrate diet. This aspect of the use of straw is not important in the context of "matching livestock systems to available resources". The following discussion emphasises the use of straw as a basal diet and the use of supplements to increase productivity.

The nutritional value of straw varies according to a number of factors, including:

- Cereal species (in general, oat, barley, sorghum and millet straw are often of higher digestibility than wheat and rice straw)
- Variety and tannin content. There is a negative correlation between tannin content and digestibility (Saini et al 1977)

- Stage of harvest (eg. wetland rice is often har-vested green)
- Length of storage (which often depends on cli-mate and storage conditions)
- Proportion of leaf to stem selected
- Fertiliser application and soil fertility
- Irrigation: straw from irrigated wheat had an in vitro digestibility of 41% compared with 34% for non-irrigated wheat (Kernan et al 1979)
- Plant diseases
- Weathering: leaf loss increases with time after removal of the grain
- Maturity.

Table. Straws from Various Cereal Crops Appear to be Highly Variable in Organic Matter Digestibility (in Vitro) (OMD). The N Content is Always Low

Wheat	28-58	0.4-1.0
Oat	34-68	0.4-1.0
Rice	40-52	0.5-1.0
Barley	34-61	0.4-1.0
Maize	31-50	0.5-1.2
Sorghum leaves	65-78	0.5-0.8

Table In vitro true organic matter digestibility (True-OMD) and lignin content of sorghum leaf and stem from 24 varieties grown under the same con-ditions. Among leaves, 80% of the variation in di-gestibility could be accounted for by the range in lignin content. Higher lignin content was associated with the presence of insoluble condensed tannins in the fi-bre fraction. Variation in digestibility of both leaf and stem fractions of sorghum straw is likely to influence productivity of livestock consuming it and should be taken into account by plant breeders.

True OMD (in vitro) (%)	Lignin (% DM)	Mean (range)
Leaf	70 (62-79)	6.0 (4.7-7.1)
Stem	65 (53-74)	6.6 (5.1-9.1)

All cereal straws have two characteristics in common:

1. Low nitrogen content
2. They are composed of cell wall components with little soluble cell contents and therefore have to be digested by microbial fermentation.

There appear to be no toxic substances in straws, except when they are mouldy. Rice straw, which forms a large proportion of the available forage in developing countries, contains high levels of oxalate and silica which are reported to be concentrated in the leaf rather than the stem. The lignin content of rice straw appears to be less than in other straws.

When straws are fed to ruminants the primary lim-itations to production are:

- The slow rate of, and low total digestibility
- The rate at which straw particles break down to a size that can leave the rumen
- The low-propionate fermentation pattern in the rumen
- The negligible content of both fermentable N and bypass protein.

TREATMENT OF STRAWS TO INCREASE DIGESTIBILITY

The productivity of animals fed straw can be in-creased by treatment of the straw to increase di-gestibility and feed intake, especially when combined with correct supplementation. The use of treated straw with supplements also allows the use of ani-mals with higher genetic merit. Methods for industrial-scale treatment of straw with alkalis (eg. ammonia and caustic soda) have been available for some time. In the developing countries the major problem with this approach is to find methods that are both acceptable and effective at the village level.

Table : Milk Production and Change in Liveweight (L Wt) in Zebu and Holstein/Zebu cows in Bangladesh on Basal Diets of Untreated or Ammoniated Rice Straw (NH_3-straw) (Urea ensiled).The Native Animals per-formed best on the Untreated Straw, but the Crossbreeds were Superior when the Degradability of the Straw was Increased by Ammoniation.

	Straw	NH_3-straw
Straw intake (g/d)		
Local	-	-
Crossbred	5.2	10.2
LWt change (g/d)		
Local Crossbred	-87	49
Milk yield (kg/98 days)		
Local	54	204
Crossbred	165	255

In some European countries, caustic soda has been used to increase the digestibility of straw for more than 40 years. A number of other techniques are be-ing applied in large enterprises (eg. ammoniation) but these methods have not been generally accepted by small farmers. Methods for treating straw with caustic soda are rarely economic, are difficult to ap-ply and are hazardous to people and animals. Thus, treating straw with caustic soda is impracticable in most developing countries.

There are a number of ways in which ammonia can be used to increase the digestibility of fibrous feeds, including methods that use ammonia gas, ammonia in solution or ammonia generated from urea. Ammoniation of straw appears to be potentially applicable to a wide range of situations. The choice of the method of ammoniation will depend on the cost and availability of ammo-nia gas relative to urea. Ammonia gas lends itself to large operations where there is the necessary infras-tructure for distribution of ammonia in cylinders or tanks. This may be readily done in oil-rich countries where ammonia gas is manufactured for agricultural purposes.

For small-scale farmers it is more convenient to gener-ate ammonia from urea by the "wet-ensiling process". Urea is a common fertiliser which is often subsidised and which farmers have become accustomed to han-dling.

It poses no hazard to human health but farm-ers are generally concerned about its possible toxicity to livestock. Ammonia is generated rapidly from urea at high temperatures when it is mixed with moist stra w, which makes the system appropriate for trop-ical but not temperate countries. A source of ure-ase may also need to be added for the more inert crop residues or fibrous byproducts (eg. bagasse; see Torres et ai. 1982). The ground seed from Jack- bean (Canavalia ensiformis) is being used commer-cially for this purpose in Colombia (Preston 1987). Urea-ensiling appears to be less effective than using ammonia gas because of the formation of ammonium carbonate, which decreases the pH of the straw (Mason et al. 1985).

Ammonia can be generated from other nitrogenous materials such as chicken manure but these tech-niques are only now being developed. The use of animal or human urine to ensile straw has been re-searched in Bangladesh and has shown promise for application on small farms. It i s, howe ver, e ssentia l t o ensure an adequate level of ur ea in the urine .

Several other chemicals can be used to increase straw digestibility including calcium oxide, acids and acid gases (sulphur dioxide) or combinations of these. All these techniques are experimental and have not yet been proved to be applicable.

One method of treatment to increase the digestibility of fibrous feeds which appears to have some applications on an industrial scale is the use of steam at high pressures. This may be feasible where this energy source is available and inexpensive (*i.e.* in sugar mills where there is usually surplus steam). The technique appears to be particularly appropriate for treatment of bagasse which is available on site at the sugar mill and where the necessary technical knowledge and equipment are also available.

MINERAL CONTENT OF STRAW

The mineral content of straws is generally low and imbalanced but deficiencies are unlikely to be manifested in animals at maintenance or working. For production of meat and milk, requirements for minerals are increased many-fold and supplements should be supplied. However, responses to mineral supplements will only occur after the major nutrient imbalances (protein and glucogenic energy) have been corrected. The mineral composition of plants depends largely on the availability of minerals in the soil and closeness to the oceans (for Na). Calcium and phosphorus contents of straw are usually below recommended levels and cobalt, copper, sulphur and sodium ma y also be deficient. The high concentrations of oxalates and silicates in rice straw also suggest that considerable amounts of calcium and magnesium salts can be lost as silicates and oxalates in urine and faeces.

DRAUGHT ANIMALS

Draught animals appear to work and survive on a wide range of fibrous diets and are able to tolerate the low digestibility of the diet and its low nitrogen content. Working bullocks in Bangladesh con-sumed greater quantities of ammoniated rice straw (ensiled with urea) than their pair mates given un-treated straw but there were no apparent differences in work output or bodyweight chan ge (Dolberg et ai. 19 81). Further evidence that working animals require little protein supplementation is the finding that young horses that were exercised grew at the same rate on a low- protein diet as those on a high-protein diet that were not exercised (Ortonetai. 1985a). Recent work suggests that excessive protein in a diet can slow the race horse (Glade 1983).

In Pakistan, Preston (personal observation) observed bullocks driving apress, crushing 400 kg of sugarcane per hour, working alternate shifts of 3 hours work followed by 3 hours for feeding and rest, and found that they were apparently able to work on this basis for 24 hours a day almostcontinuously for 6 months on a diet of only sugarcane tops. In contrast, a diet of sugarcane pith of higher digestibility (70%), supplemented with urea and minerals barely supported maintenance in 'growing' steers (Preston et. al. 1976). The exclusive use of sugarcane tops as feed for working oxen is common practice on sugar estates with a priority for employing people. Obviously the influence of exercise on nutrient requirements is a subject requiring a great deal of research.

GROWING ANIMALS

Moderate rates of liveweight gain can be obtained with young ruminants on diets based on crop residues provided that a number of principles areapplied. These are discussed below.

Green Forage

A small quantity of highly digestible green forage ap-pears to have beneficial effects on rumen function on diets based on crop residues. The results in Table indicate that such improvements in the rumen ecosystem carry through to improved animal perfor-mance. Azolla pinnata, a water plant which grows symbiotically with the N-fixing alga Anabaena azolla, appeared to be even more effective than a mixed concentrate meal in promoting live weight gain in cattle fed a diet of wheat straw and sugarcane tops.

Table: Small amounts of the water plant Azolla pinnata appear to be as effective as a balanced concen-trate in improving the utilisation of a diet forcattle based on crop residues.

	Concentrate	Azolla
LWt gain (g/d)	140	330
Feed intake (kg DM/d)		
Wheat straw+sugarcane tops		
Supplement		
Total	3.6	3.0
Feed conversion(kg DM/kg gain)	26	9
OMD (%)	45	59

The advantage of using the foliages of legume trees as "green" supplements is that these are rich in protein (25 to 30% in dry matter), at least part of which appears to escape rumen fermentation. According to Bamualim et al. (1984)' leucaena leaf meal was as effective in stimulating voluntary intake of a low-N hay as was case in infused into the abomasum.

The data in Table and Figure confirm the effectiveness of leucaena foliage as a supplement for cattle fed basal diets of rice straw. In the feedingtrial in Colombia, which involved 36 weaner steers fed ad libitum ammoniated rice straw for 90 days, fresh leucaena leaves (2kg/d) were as effectiveas 500g/ dohice polishings in stimulating live weight gain (from 200 g /d in unsupplemented animals to 550g/d with the supplements).

Table: Leucaena Foliage (+LF) Increased Dry Matter Intake and Nitrogen Retention in Cattle given Diets Based on Untreated or Treated (sodium hydrox-ide) Rice Straw (NaOH-straw)

	Untreated straw		NaOH straw	
	–LF	+LF	–LF	+LF
DM intake, kg/d	5.1	6.6	3.6	3.9
DM digestion,	38	42	47	48
N retention, /d	–6	+7	– 6	+26

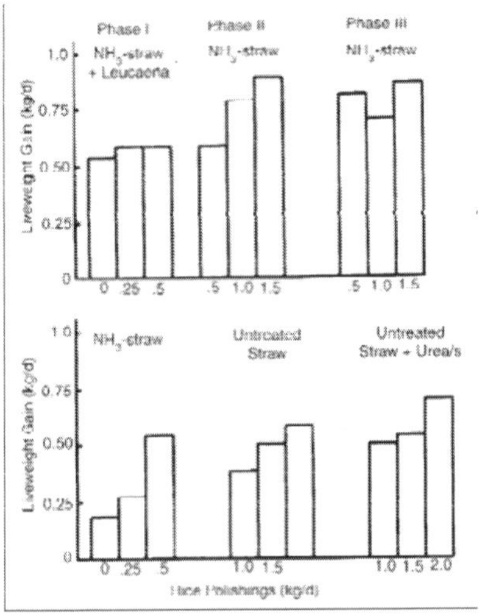

Fig. Development of a Cattle Feeding System Based on Ammoniated Rice Straw.

In phase 1, the basal diet of 6 groups of weaner bulls was ammo- niated straw (ammonia gas); the supplements were rice polishings at 0, 250 and 500g/d in the pres-ence or absence of fresh leucaena folzage (2kg/d). In phase 2, the treatments were ammoniated and un-treated rice straw with different levels of rice polishings. In phase 3, the treatments were ammoniated and untreated straw, the latter was supplemented with urea/S Lucerne hay comprising 13% of a diet based on maize cobs increased liveweight gain of cattle by 100% when the basal diet was untreated and by 50% when the maize cobs were treated with ammonia.

Bypass nutrients and glucogenic compounds. The data in Table are from experiments in Australia with sheep fed oatstraw.

Table. Formaldehyde-casein (Formal-C), Rapeseed Meal (RSM) and Sunflower Seed Meal (SFM) (both Meals with and without Formaldehyde Treatment) Increased Fibre Digestibility, Dry Matter Intake, Liveweight Gain and Wool Growth in Sheep Fed a Basal Diet of Oat Straw.

	Urea	Formal-C RSM		Formal- SFM		Formal-SFM
DM intake (g/d)	1316	1779	1750	1899	1909	1977
Digestibility (%)						
O M	45	50	48	46	48	46
NDF	40	47	45	45	44	45
LWt change (g/d)	30	132	140	144	144	157
Wool growth (g/d)	4.5	11.0	10.5	10.1	10.1	11

The two feeding trials where the pattern of response to the supplement was measured. The results from the trial in Bangladesh are particularly interesting as they show that a daily supplement of only 50g of fish meal to cattle tripled productivity on a diet ofammoniated rice straw.

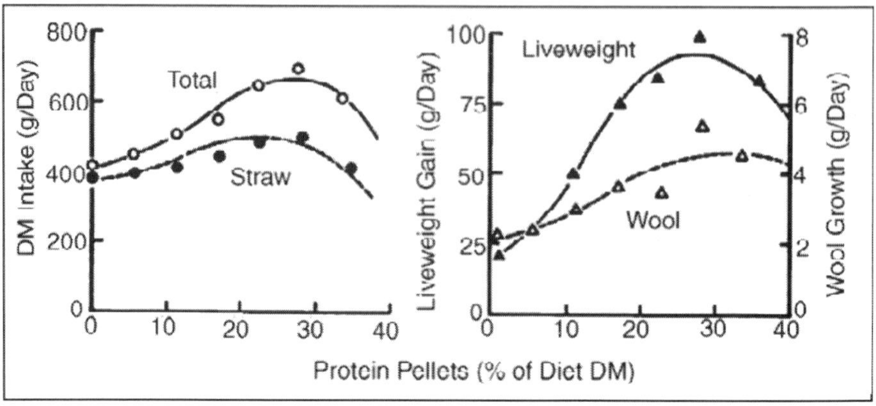

Fig. Feed Intake, Wool Growth and Liveweight Gain of Lambs given Barley Straw/urea and a Bypass Protein Pellet (Cottonseed Meal 80%, Meat Meal 8%, Soya Bean Meal 10%,Minerals 2%).

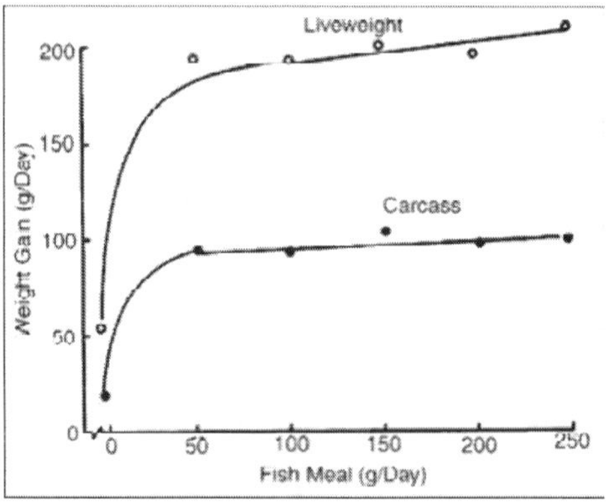

Fig. A Small Supplement of Fish Meal Dra-matically Increased g1'Owth in Live and Carcass Weight in Young Cattle Fed a Basal Diet of Ammoniated (urea-ensiled) Rice Straw in Bangladesh.

Recent research in Thailand and Australia provides strong support for the concept that the critical supplementary nutrients on a straw-based diet are bypass protein, starch and long-chain fatty acids. High rates of growth were obtained when the ammoniatedstraw (urea ensiling in Thailand and ammonia gas in Australia) was supplemented with starch, protein and oil in byproduct meals that are known to escape rumen fermentation .

Table: Effects of Supplementation of Ammoniated Rice Straw with a Mixture of Fat, Protein and Rice Starch. Young Brahman Bulls (36) Weighing150kg Liveweight were Used in an Experiment Lasting 153 Days.

Supplement*, kg/d			1 2 3
Feed intake (kg/d)			
Rice straw	5.81	6.22	4.35
Supplement	0.87	1.74	2.62
Total	6.68	7.96	6.97
LWt gain (kg/d)	0.47	0.84	0.93
Feed conversion			
(kg DM/kg gain)	14	9	7.5

Table: The Effects of Various Levels of By-pass-protein (Largely Cottonseed Meal) Supplement on the Live Weight Change of Cattle (320kg Liveweight) given a Diet of Ammonia-treated or Untreated Rice Straw, O.5kg Molasses/Block (15% Urea) to Provide Fermentable N, and O.6kg Rice Polishmgsto Supply Small Amounts of Starch and Lipid

Straw Preparation	Protein meal	LWt gain (kg/d) (g/d)
None	0	38
	0.4	365
	0.8	292
	1.2	306
Treated with 3% NH3gas	0	236
	0.4	497
	0.8	601
	1.2	639

The interaction between ammonia treatment of straw and response to bypass nutrients is illus-trated in Figure which summarises the perfor-mance of steers fed on ammoniated or untreated rice straw. During phase 2 (approximately 260-350kg liveweight), there were linear responses in growth rateto a supplement of rice polishings on both ammoni-ated and untreated straw. However, 2kg/d of the sup-plement was needed when untreated straw wasgiven compared with only 500g/d to support the same rate of liveweight gain on ammoniated straw. The inter-action continued to be manifested duringthe final finishing phase of the cattle (350-450kg liveweight), when urea and sulphur were sprinkled on the untreated straw.1t can be concluded that ammoniation raised the availability of nutrients from straw and also led to a more balanced array of nutrients to the cat-tle since, lower levels of dietary bypass nutrients were needed to optimise performance.

LONG-CHAIN FATTY ACIDS

Evidence which demonstrates the benefits of having a dietary source oflong chain fatty acids in straw-based diets is summarised in Figure. Lambs fed ammoniated wheat straw and minerals increased their growth rate and feed conversion in response to di-etary long chain fatty acids given in the form of in- soluble calcium soaps (Ca-LCFA). The response was only apparent when bypass protein was also given. Subsequent work confirmed theinteraction between LCFA and bypass protein, and showed significant increases in carcass fatness in response to the LCFA supplement (given as long-chain fatty acid prills).

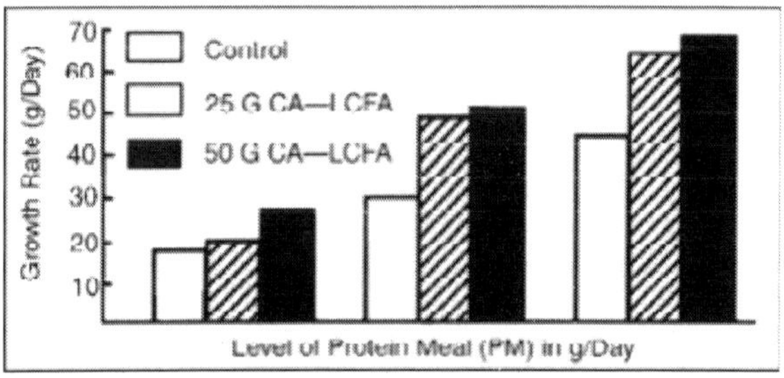

Fig. A Supplement of Long-chainfatiy Acids (as Insoluble Calcium Soaps) Increased the Growth Rate of Lambs Fed Ammoniated Wheat Straw and Bypass Protein

From the discussion in earlier chapters it is obvious that fibrous crop residues can only support low milk yields in ruminants because of their inability tosup-ply enough protein and glucogenic energy to balance the VFA energy. These feeds are the most, and often the only, available resource on manysmall farms ill developing countries. Therefore every attempt should be made to use them as efficiently as possible. Milk synthesis represents a greater drain on critical nutri-ents than any other physiological state.

In view of the nutritional limitations of most crop residues, set by low digestibility and low N content, and the high demand for amino acids, long-chainfatty acids and glucogenic compounds, milking an-imals must be given high priority for:

- The available supplements and
- The residues that have higher potential digestibility or that have been treated to improve digestibility.

The scientific basis for feeding milking animals on crop residues is discussed in the following section.

Effect of Ammoniation

Upgrading the nutritional value of straws by ammoniation (urea ensiling) is more likely to be economi-cally justifiable for lactating animals than foranimals in other physiological states. In most circumstances the sale value of the extra milk produced should more than cover the cost of processing thestraw.

Table. Milk Yields and Changes in Live Weight (L Wt) in Cattle and Buffaloes Fed Basal Diets of Un-treated and Ammoniated (Urea-ensiled) Rice Straw (NH3-straw)

Straw	Milk yield (kg/d) NH_3-straw		LWt change (g/d)	Straw NH^3-straw
Cow1	1.1	2.3	–149	+109
Cow2	2.4 (4.6)	3.4 (4.9)	–266	+93
Buffalo3	2.4 (6.8)	3.2(7.6)	–17	+93

The data in Table summarise research from Bangladesh and Sri Lanka where ammoniated or un-treated rice straw was the basis of the diet for Zebuand buffalo cows. The cattle responded significantly to ammoniation of the straw, both in milk yield and in liveweight change. Ammoniation of the straw in-creased feed intake, and obviously reduced the need for bypass nutrients since, the amount of milk produced per unit of concentrate fed wasinc rea sed by an a verage of 53%. The proportion of the total diet represented by concentrates was reduced from an average of 14% to 9% due to increased intake of the treated straw.

Table: Ammoniation of Rice Straw (NH3-straw) permits Reductionin Amount of Concentrate Required for Milk Production

Straw	NH3 straw	
Concentrate (% diet DM)		
Zebu x Friesian	11	8
Zebu	21	14
Buffalo	11	5
Milk produced (kg/kg concentrate)		
Zebu x Friesian	1.7	2.7
Zebu	2.8	4.2
Buffalo	4.0	6.0

Effect of Green Forage

The effect of green forage, in the form of gliricidia leaves (Gliricidia sepium), is illustrated in Figure. On a diet of untreated straw, feeding gliricidia foliage at approximately 15% of the dietary dry matter in-creased milk yield by 22%. With ammoniated straw as the basal diet gliricidia comprised 10%of the diet and milk yield was increased by 14%.

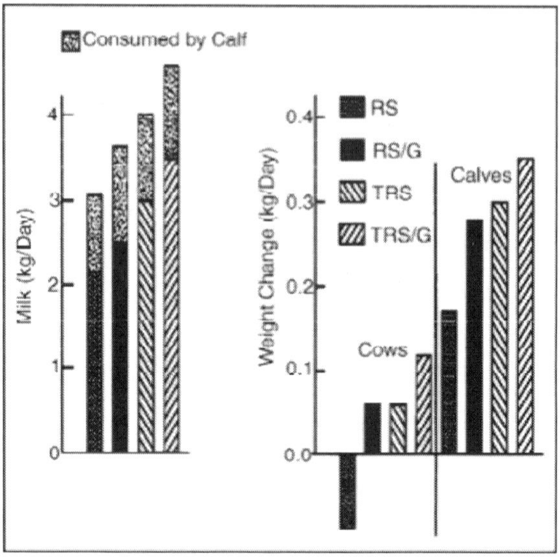

Fig. Supplementation with Fresh Gliricidia foHage (G) or Ammoniation of Straw (Urea Ensiling) (TRS)increased Milk Yield and Body Weight Gain of Buffalo Cows and Calves Fed a Basal Diet of Rice Straw and 1 kg/day of Concentrate (RS). Best Results were Obtained when Straw was Treated with Ammonia and Supplemented with Gliricidia Goliage (TRS/G)

Bypass Nutrients

The data in Figure and illustrate the results of trials in which responses different amounts of protein supplements were measured. In Bangladesh there was a linear increase in milk yield in Zebu cows when fish meal was given as a supplement to a basal diet of ammoniated rice straw.Milk yield was increased by 23%, fat percentage by 8% and liveweight gain by 110% when 1000g of coconut cake were fed daily to lactating buffaloes in Sri Lanka on a basal diet of ammoniated rice straw and minerals.

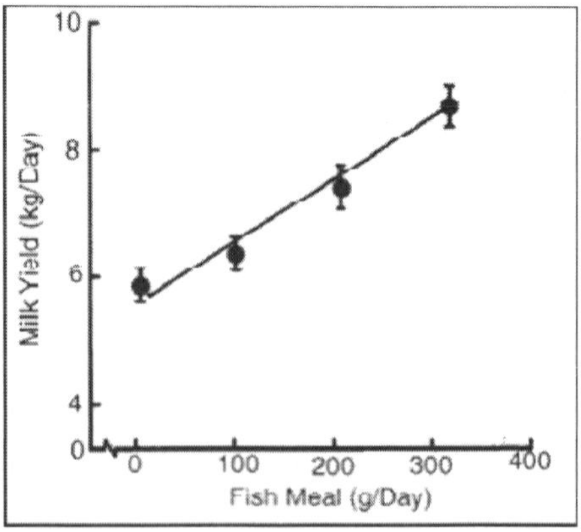

Fig. Milk Yield of Native and Crossbred Cattle Fed Ammoniated (Urea-ensiled) Rice Straw in Bangladesh was Increased Linearly with. Small Amounts of Fish. Meal (0 to 400g/d).

Table. Supplementing Ammoniated (Ureaen-Siled) Rice Straw with. Coconut Cake Increasedthe Milk Yield and Liveweight Gain of Buffaloes inS7'i Lanka.

Coconut cake, kg/d	0	1
Milk yield (kg/d)	2.6	3.2
Fat (%)	9.1	9.8
LW gain (kg/d)	0.1	0.21

WOOL GROWTH

The effects of strategic supplementation on wool growth in sheep are illustrated in Figure. Wool growth was increased when either a protein meal(cottonseed cake) or a readily-digestible forage (lucerne hay) was the supplement in a wheat-straw diet provided with fermentable N by adding urea or by ammoniation of the straw.

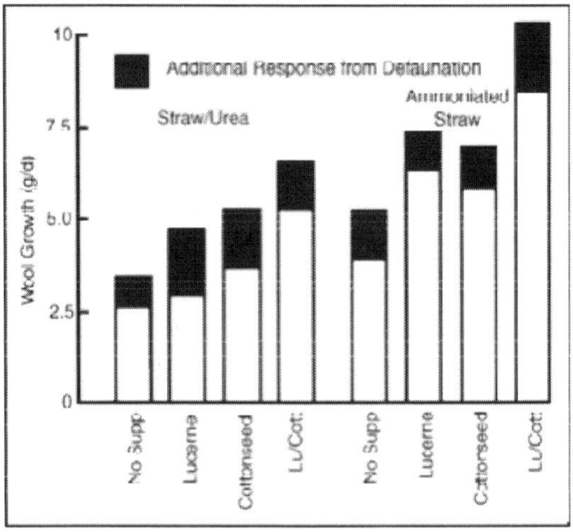

Fig. Supplements that Provided Rumen "Acti-vators" (Lucerne Hay) and/orbypass Protein (Cotton-seed meal) Increased the Wool g1'Owth of Sheep Fed a Basaldiet of Wheat Straw that had been Sprayed with Urea or Ammoniated.

There wereadditional benefits when the sheep were defaunated (Source: S H Bird, B Romuloand R A Leng, unpublished data.)

The combination of the two supplements gave the best results. Responses were more marked when the straw digestibility was increased by ammoniation and when the sheep were defaunated.

The data presented previously suggest that the optimal wool growth response in growing lambs fed a barley straw/urea basal diet occurredwhen the protein meal (80% cottonseed meal, 10% soya bean meal, 8% meat meal and 2% minerals) comprised about 30% of the diet dry matter.

METABOLIC DISORDERS ASSOCIATED WITH AMMONIATION OF FEEDS

Adding ammonia to ruminant feeds to provide fermentable N has been researched for a number of years. In early trials, ammoniation of molasses washighly detrimental because the ammonia gas reacted with the sugars, increasing the temperature and ap-parently resulting in formation of toxic methyl im-idazole compounds which induced a severe nervous disorder (Tillman et al. 1957, Bartlett and Broster 1958).

In countries with a temperate climate, ammonia-tion of straw (to increase digestibility) requires 3-6 weeks (versus 10 days or less in the tropics) because of low ambient temperatures. In order to reduce the treatment time, methods have been developed in which the straw is heated. Large ovensare manufac- tured in Europe to treat several tonnes of straw at 90°C, allowing the treatment time to be reduced to one day.

This practice has been commercialised in Europe and ammoniated straws have been fed to livestock with no reported ill-effects (H Sundstol, personal communication). However, ammoniated hays caused bovine hysteria when fed to cattle (for review see LaBore et al. 1984). The problem also occurs occasionally with ammoniated straw. This effect may be related to the proportion of the treated forage or straw in the animal's diet, since, recent research in Australia (with one batch of rice straw and one of wheat straw) indicated that bovinehysteria developed whenever the treated straw comprised a large proportion (70-80%) of the diet.

In studies by Perdok and Leng (1985, 1987), in which 64 yearling cattle were given thermo-ammoniated rice straw from a failed crop, almost all theanimals developed hyperexcitability, with the fol-lowing symptoms: animals were restless and blinked rapidly, their pupils dilated and their vision was apparently impaired; involuntary twitching, trembling, loss of balance and frequent urination and defecation were observed. In addition, respiration rate was high, heart rate was low and the animals salivated copiously; they bellowed and perspired. The most obvious and most dangerous symptom wassudden stampeding: the cattle galloped in circles, were inclined to collide with each other and also to run into fences. Often the animals galloped at suchspeed that they broke limbs, and in one instance an animal died. The symptoms usually lasted for 5 minutes and were repeated at 20- to 30-minute intervals. The affected animals appeared normal between these attacks and tended to return to feed on the treated straw. The condition was alsoinduced in animals fed the same rice straw after it had been treated with ammonia under plastic sheets for 4 weeks.

The syndrome has occurred also in young calves suckled by cows consuming a diet containing a ma-jor proportion of ammoniated rice straw: both cows and calves showed symptoms. Pasteurised milk from these cows also caused the hysteria when fed to calves. Under some circumstances calves suckled by cows fed ammoniated hays have died. These observations indicate that the toxic compound(s) is transmitted in milk and is unaffected by pasteurisation.

Bovine hysteria in cattle fed straw has not so far been reported from Northern Europe, possibly be-cause the proportion of ammoniated straw fed rarely exceeds 30% of the total diet in intensive livestock feeding systems. A recent report from Southern Spain (Cabrera et al. (1987) has, however, shown that in warm climates feeding ammoniated straw to cattle and sheep can be extremely dangerous. Cabrera et al. (1987) reported numerouscases of hyper-excitability and death in sheep and cattle fed straw when this was initially treated with ammonia gas af-ter 1100 hours and on a warm day in summer.

Ammoniation through urea ensiling is highly un-likely to result in toxic compounds being produced, particularly if the moisture content of the straw iskept high. It is advisable, however, always to pre-pare straw when the temperature is low.

The transmission of toxic compounds in milk from cows fed ammoniated forages suggests that care should be taken in feeding ammoniated forages tomilking animals.

FIBROUS AGRO-INDUSTRIAL BYPRODUCTS [MH]

There are a number of fibrous residues that, untreated, have only limited application as the basal component in livestock feeds (eg. bagasse, palm- pressed fibre, cocoa pods and rice hulls). The limitation to all these feed resources is their extremely low digestibility and, even when supplemented with essential nutrients, intake is too low to support maintenance. They are, how ever, often used as fillers and sources of roughage in high-concentrate diets.

Most of these residues arise from industrial processing and are therefore available in large quantities at the factory site. The concentration of these byproducts at the factories is an incentive to finding ways to use them as ruminant feeds by using 'relatively sophisticated' technology to increase their digestibility. Techniques such as briquetting, steam treatment and ammoniation can be considered under these circumstances.

The factories where these residues are produced usually have the infrastructure necessary to enable adequate servicing of the machinery that is normally required for any large-scale industrial treatment of a fibrous residue. Often the residue is used as fuel in the factory (eg. bagasse, palm-pressed fibre and peanut hulls), but frequently there are surpluses. Little research has been done on these residues, with the ex-ception of steam-treated bagasse which is now being used commercially in diets for cattle in both Brazil (E L Caielli, personal communication) and Colombia (Preston 1987).

Some of the original research on steam treatment of sugarcane bagasse was done in Mauritius (Wong et al. 1974). Treating the bagasse with high pressure (14kg/cm2) steam for 5 minutes raised dry matter digestibility from 28 to 60% (rumen nylon bag method; 48 hr incubation). Early attempts to use the treated bagasse as the basis of the diet for growing cattle focussed attention on the need to supplement with bypass protein (fish meal), LCFA and glucogenic pre-cursors (maize grain).

Table: Calves lose weight on a diet of steam-hydrolysed bagasse (200°C for 10minutes) supplemented with only urea and minerals. Significant improvements in growth and feed conversion wr'e brought about when fish meal (bypass protein) and/or maize grain (glucogenic energy and oil) were added to the diet.

	Control		**Fish meal (1kg/d)**	**Maize meal (0.25kg +1kg/d)**	**Fish meal 1 (0.25kg/d)**
Bagasse intake(kg DM/d)	4.0	4.4	4.5	4.5	
LW change (kg/d)	–0.17	+0.08	+0.16	+0.33	
Feed conversion	–	34	23	12	(kg DM/kg gain)

The data in Figure show how the commercial feeding system was developed in Colombia, starting from the premise that the treated bagasse should con-tribute the major part of the diet and that locally available supplements should be used. The diet which gave the best results (810g/d of liveweight gain), and which was chosen for the commercial programme, contained (% dry matter basis): 53 steamed bagasse, 16 final molasses, 2 urea, 15 gliricidia foliage, 6 poultry litter, 7 rice polishings and 1 salt.

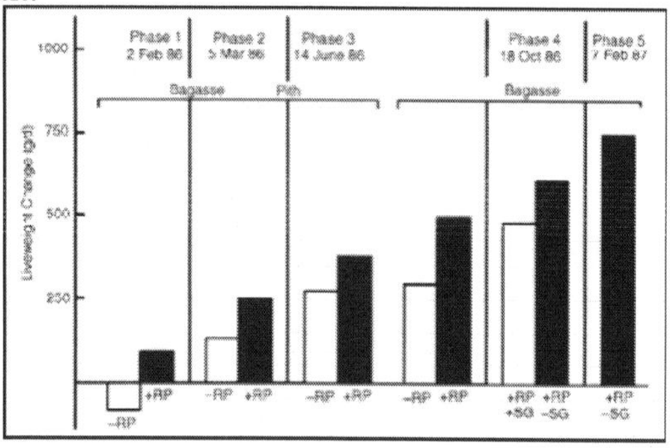

Fig. Development of a Cattle Feeding System Based on Hydrolysed Cane Bagasse (Steam at 14kg/cm2 for 5 Minutes).

The same 8 groups of animals (weaned bulls of 180kg initial liveweight) were used throughout the twelve month trial. In phases 1 and 2, the basal diet wastreated bagasse pith supplemented daily with molasses/urea (10% urea), gliricidia foliage (2%of liveweight daily) and poultry litter (0.2% of liveweight). The variable was the presence or absence of rice polishings (0.2% of liveweight, RP). In phase 3 the treatments were untreated bagasse and bagasse pith, with or without the rice polishings. In phase 4 the effect of additional long fibre was studied (from African Star grass S G) with the basal diet using treated bagasse and with the addition of rice polishings. In phase 5, all the animals were given the best diet based on the results obtained in the previous phases

The molasses was used as a carrier for urea and also provided trace elements; gliricidia foliage was the source of fibre, vitamin A and some fermentable and bypass protein; poultry litter provided fermentable N and macro (Ca, P, S) and micro (Cu, Co, Zn etc) minerals and rice polishings was the sourceof nutri-ents which improved the balance of amino acids and glucose available in the total nutrients absorbed and this increased the efficiency of feed utilisation. The lipid content of rice polishings was expected to be used highly efficiently in this otherwise low fat diet.

FRESHLY HARVESTED GRASSES

Elephant (or Napier or King) grass (Pennisetum pur-pureum), guinea grass (Panicum maximum) and sugarcane (Saccharum officina rum) are among the high-est yielding perennial crops, in terms of total biomass production per unit area and efficiency of solar en-ergy capture. Pennisetumspecies and sugarcane are used widely in the American tropics as a reserve crop ("ensilaje vivo") for feeding during the dry season. Sugarcane has the advantage that its energy value as feed increases as it matures, since, the accumulation of sucrose more than compensates for the increasing lignificationof the cell wall. The "stress" of the dry season (low soil moisture, low soil N and [generally] lower air temperatures) also stimulates the storage of sucrose in the stem.

Pennisetum yields most biomass when it is har-vested at 6-month intervals, although it is well understood that nutritive value is highest when harvesting is over shorter intervals (about 6 weeks). On the farm, it is difficult to maintain the rigorous practice of fertiliser application, irrigation and fre- quent cutting and as a result the N content and dry matter digestibility of the infrequently harvested for-age are low.

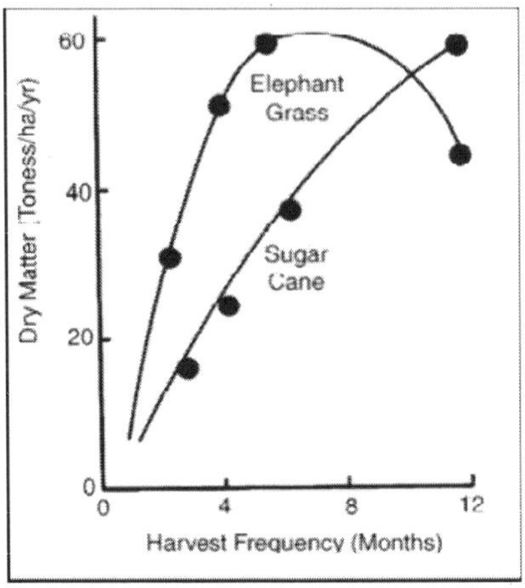

Fig. Annual Biomass Yields from Sugar Cane and Elephant Grass Harvested at Different Intervals

Early work with these forages emphasised their use as "roughage" supplements in concentrate diets for intensively managed (usually confined) milking and fattening cattle and buffaloes. This policy predicated against the efficient use of these feed resources since, the drop in rumen pH when cereal grain is fed depresses the digestion of fibre.

In the following section examples are given of re-cent research in which these forages have been the basis of the diet and have been strategically supple-mented.

Legume Forages

In Costa Rica , lactating goats were fed a basal diet of King grass (Pennisetum purpureum) supplemented with a fixed allowance of green banana fruit and in-creasing amounts of the leaves of the legume tree Erythrina poeppigiana. Total dry matter intake and milk production increased linearly with legume sup-plementation. There was only a mini-mal substitution of the King grass in the diet (intake fell from 690 to 600g dry matter / day).

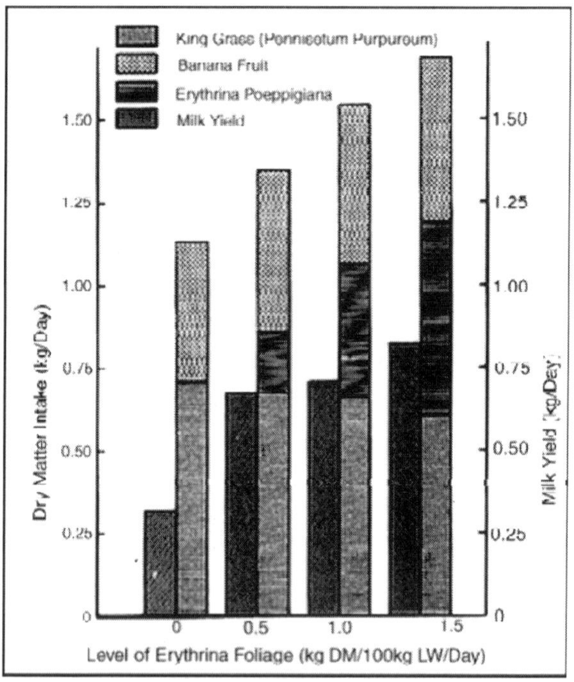

Fig. Effect on Dry Matter Intake and Milk Yield of Goats of Supplementing their Basal Diet of King Grass and Banana Fruit with Foliage of Erythrina poep-pigiana.

In research in Colombia the foliage of the legume tree, Gliricidia sepium, was given as a supplement to weaned steers fed a basal diet of freshly har- vested King grass during the dry season. Growth rate increased curvilinearly in response to in-creasing levels of the legume foliage, with the opti-mum legume content of the diet being about 30%.

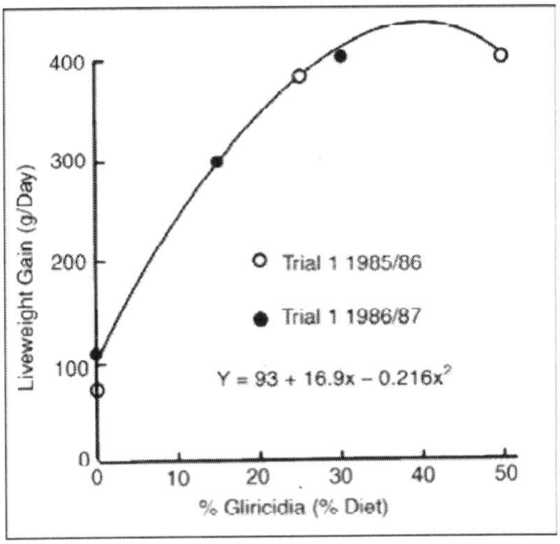

Fig. Effect of Increasing Levels of Gliricidia Foliage on the Growth Rate of Weaned Bulls (Initial Weight 180kg; 4 Animals per group) given a Basal Diet of King Grass

Similar effects have been reported with West African Dwarf goats in Nigeria fed combinations of Guinea grass (Panicum maximum) and gliricidia (M. van Houtert, personal communication).

Feeding a supplement of gliricidia to growing sheep on a basal diet of fres hly h arve sted Brach ia ria mu lti-formis ga ve a signifi ca nt respons e inbodyweight gain and in wool growth when the legume comprised up to 28% of the diet.

Table: Effect of Supplements of Leaves from Gli-Ricidiamaculata on Performance of Growing Sheep Fed a basal diet of Brachiaria Mileformis.

% Gliricidia in diet (DM basis)				
	0	28	50	65
LW gain, g/d	25	43	44	39
DM intake, g/d	530	610	690	690
DM conversion	21	14	16	18
Carcass, kg	10	12	12	12
Clean fleece, g	280	350	320	360
Staple length, cm	4.3	4.7	5.4	5.5

Results of studies with milking cows fed diets based on freshly harvested Setaria grass and given a range of supplements are shown in Table Themost appropriate supplement was groundnut cake since, it maintained body weight and supported the same lev-els of milk production in cows as cerealgrain or molasses-based supplements fed at twice the rate. The implication is that the principal deficiency in the absorbed nutrients was amino acids and this was effectively corrected by the ground nut supplement, which provided bypass protein.

Table: Effects of supplements based on molasses, cereal brans/oil cake or groundnut cake onmilk yield and liveweight change of Friesian cows given a basal diet of freshly harvested grass (Setaria kazangula) ad libi-tum.

	No suppl.	Cereal/oil cake	Molasses/urea cake	Groundnut cake
Milk yield, kg/d*	6.1	7.6	6.8	7.9
LW change, kg/d	–0.72	0.0	–0.83	0.13

SUGAR CANE

Growing sugarcane, more than any other crop, max-imises the yield of biomass per unit area.

Sugarcane is an ideal crop to optimise biomass utilisationbecause:

- Its C-4 pathwa y for photosynthesis confers both high-yield potential and efficient capture of solar energy
- The requirements of the large-scale sugar indus-try have promoted agronomic characteristics in the plant which facilitate fractionation
- Its perennial growth habit and disease resistance aid maintenance of soil fertility and reduce soil erosion, allowing monocultural practices with only minimal use of inputs derived from fossil fuels. There are a number of sugarcane byproducts that can be used in animal f e e ds. Deta ils of th e tw o pri n c i -p a l me th ods of e x tr a c ti ng sucrose fromsugarcane and the associated crop residues and resultant byproducts are outlined in Figure. Of the two extraction processes, the industrial technology produces the most byproducts, the principal one being molasses. In the artisan (small-scale) system for production of "Gur" (Indian continent) or "Panela" (Latin America) there is no centrifugation and therefore no final molasses.

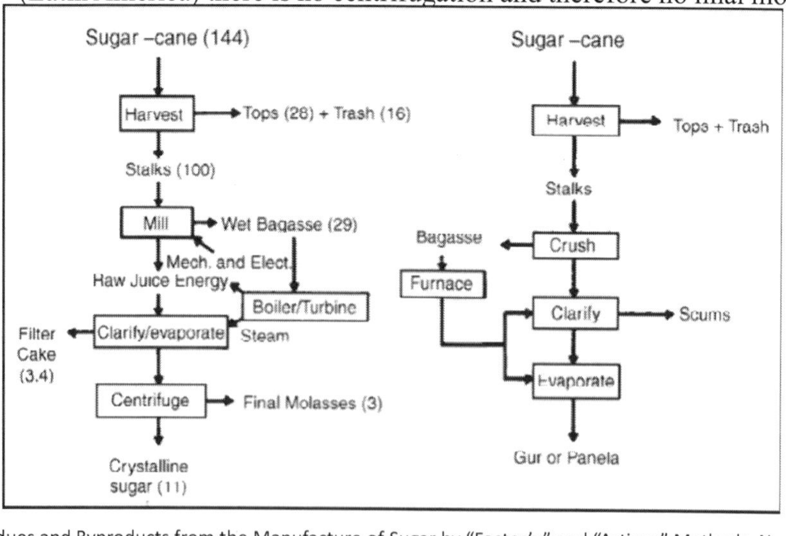

Fig. Residues and Byproducts from the Manufacture of Sugar by "Factor'y" and "Artisan" Methods. Numbers in Brackets Indicate Approximate Quantities (fresh basis) Relative to Cane Stalk = 100. I

WHOLE SUGAR CANE

Urea Supplementation

In all the trials discussed here, urea was added to the sugarcane at levels that would satisfy the needs of the rumenmicro-organisms for fermentable nitrogen (approximately 3% of the dry matter of the diet).

The optimum level was validated in an experiment in which urea concentrations were varied from zero to 4% of the sugarcane dry matter. All cattlewere supplemented with rice polishings (1 kg/day) and minerals. All the parameters of animal performance increased curvilinearly with increasing urea in the diet.

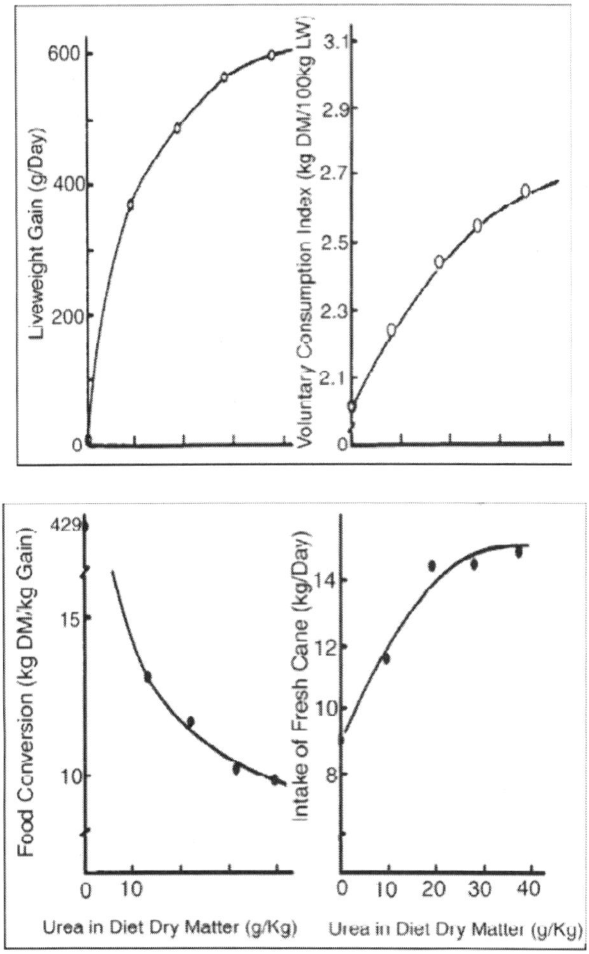

Fig. The Relationship between the Level of Urea in the Diet and Live Weight Gain, Feed Intake and Feed Conversion of Cattle a Diet of Sugar Cane and 1 kg/d of Rice Polishings (Alvarez and Preston 1976b)

When rice polishings were absent from the diet, sugarcane intake remained low and the animals lost weight irrespective of urea level; the only positive response was indigestibility (Ferreiro et al. 1977).

Cattle growth rates and feed conversions were similar when a urea solution was sprayed on to the cane (as an aqueous solution or in dilute molasses), or was given in a separate feeder as concentrated (10%) solution in molasses.

Foliage from food crops and legume trees

Cassava tops and leucaena forage given as supplements to cattle fed adlibitum chopped whole sugarcane plus urea, increased the cattle's voluntary intake of the total diet but reduced the intake of sugarcane (Meyreles et al. 1977; Hulman and Preston 1981); improvements in growth rate were small (from -40 to +140 g/day with cassava as the principal supplement; and from 60 to 200g/day when leucaena was fed).

A supplement of sweet-potato foliage added to a basal diet of chopped sugarcane and urea did not depress intake of the cane and increased total dry matter consumption by 34% (Meyreles and Preston 1978a). Liveweight gain was n ot mea sured in this tria l but s ign ifican t improvemen ts in liveweightgain of cat tle were observed when this forage was added to a ration of derinded cane stalk (cane pith).

This indicates that the foliageof sweet-potato is one of the most suitable "protein-rich" forages to use with sugarcane. The reason may be its higher rate of degradability in the rumencompared with ei-ther cassava or leucaena forage (Santana and Hovell 1979a, b), which results in a low rumen "load" while at the same time it stimulates rumen function.

Fresh banana leaves, which are only slowly digested in the rumen, were particularly unsuitable as a supplement as they reduced the intake of sugarcaneand resulted in a lower total dry-matter intake (Meyreles and Preston 1978b).

A comprehensive series of experiments with sugar-cane as a basal feed for cattle was carried out in Mexico and the Dominican Republic with the aim of understanding the constraints associated with feeding this crop as the basis of the diet for growing/fattening and lactating cattle (Preston and Leng1978a,b).

The supplement that promoted the highest level of productivity was rice polishings given at a level of 10-15% of the diet dry matter. Similar conclusions were reported by Creek et al. (1976), who fed diets of 59% sugarcane (dry matter basis) supplemented with rice polishings and cotton seed meal to Boran (Zebu) cattle in Kenya.

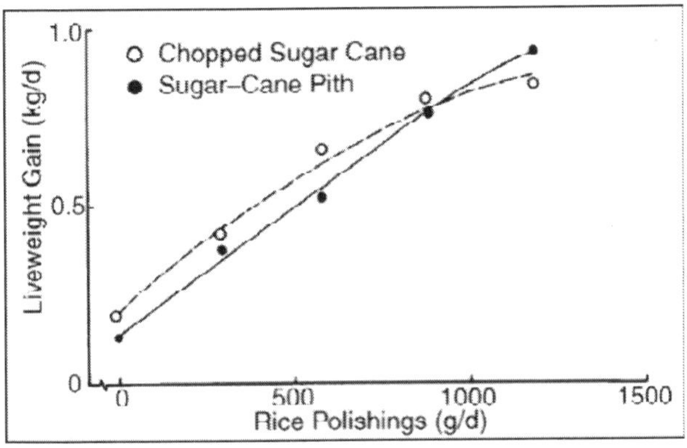

Fig. Effects of Supplementation with Rice Polishings (which Supplies Bypass Protein, Bypass Starch and oil) on Growth Tates of Zebu Bulls in Mexico Fat-tened on Basal Diets of Whole Sugarcane which has been Chopped or Derinded by the "Tilby" Separator Process

Rice polishings are relatively rich in protein(amino acids), lipid and starch, which are the critical nutri-ents that need to be adsorbed where the basal feed is digested solely by rumen fermentation. Experiments with cattle fitted w i th du ode n al ca n nu la e s h ow ed tha t the gr ea t er pa rt of the s tar ch (brokengrain) in the rice polishings bypassed rumen fermentation (Elliott et al. 1978a); the amounts of microbial and dietary non-ammonia nitrogen arriving atthe duodenum also increased in direct proportion to the amount of rice polishings in the diet (Elliott et al. 1978b) indicat-ing that rumen bacterial growthwas stimulated and also that protein in rice polishings was protected from rumen fermentation.

The development of a sugarcane-based feeding sys-tem for dual purpose (Brown Swiss x Zebu) cows and their calves was described by Alvarez and Preston (1976b) and Alvarez et al. (1977, 1978). Best re-sults were obtained w ith co mb in ed s uppleme nt atio n of 500g of rice polis hings pe r da y andrestricted graz-ing (3 hours/day) on a leucaena "protein bank" (Al-varez et al. 1978). Complete substitution of leucaena for the rice polishings led tolower milk yields and loss of bodyweight in the cows which was attributed to mimosine toxicity (Alvarez and Preston 1976a) but may have been also aresult of a lower fat intake.

Sugarcane tops are the traditional feed for draught animals (cattle, buffaloes and mules) employed in the harvesting of cane and its transport to the sugar mil. The utilisation of cane tops by work-ing animals has apparently not been studied. Cattle appear to maintain bodyweight while carrying out quite arduous work on a diet of only sugarcane tops. It appears that fermentative digestion of cane tops in the rumen provides an adequatebalance of nu-trients for maintaining bodyweight when the energy (acetate) available (relative to protein) is reduced by work. Implicit inthis statement is the concept that work increases feed intake.

Table : Composition of Sugarcane Tops and of the Rind and Pith Fractions Produced by the "Tilby" Process.

Pith	Rind		
Dry matter (DM) (%)	22	39	27
Composition (% DM)			
Protein (N×6.25)	1.4	3.2	2.7
Ether extract	0.2	1.0	0.8
Total sugars	46	24	27
Fibre	45	70	57
Ash	1.9	3.1	5.3
Sulphur	0.2	0.3	0.4

The apparent high nutritive value of cane tops for draught animals contrasts with their inability to sup-port growth in young animals without supplementation. The data summarised in Figure show that cane tops support high growth rates of cattle only when appropriately supplemented.Growing Zebu steers given a basal diet of chopped sugarcane tops grew at almost 700g daily when this feed was supple-mented with urea and 1kg of rice polishings per day. Liveweight gain was the same on chopped cane tops as on chopped cane stalk. However, feed utilisation efficiency was higheron the cane stalk, apparently because of its higher content of solu ble sugars, which could have led to a larger proportion of propionate in the rumen VFAs.

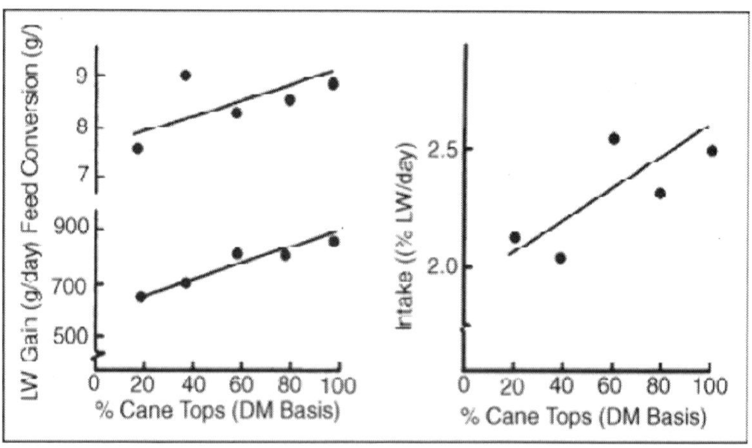

Fig. In a Fattening Diet f01' Cattle, Increasing the Proportion of Cane Tops (Replacingchopped Cane)

The constraints to the wider use of sugarcane tops in livestock feeding a r e th e ec on omi c s of h a rves ting an d tran sporti n g th e vol u minous material. Tradi-tionally, sugarcane tops have been collected by small-holder farmers using draught animals, and even as "head loads". The increasing practice of burning the cane prior to mechanical, and even hand harvesting has reduced considerably the quantity of cane tops available.

Cane tops are one of the most under-utilised re-sources even in those countries in which shortages of animal feed and of fuel are most evident. When they are fed to livestock, however, they have generally been used in efficie ntly because of the lack of kn owledge of the need for "key" supplements. If they could be treated safely (and economically) to increase fibre digestibility this would improve their feeding value. Urea-ensiling has been effective in this respec.t (A. Boodo, unpublished).

In the late 1960s a technology (Tilby Separator Process-Figure) was developed for separating the cane rind from the pith. The aim was to use thecane rind in the manufacture of compressed boards which could substitute for plywoods that are usually imported into tropical countries. The residual pith was to be used for sugar extraction or alcohol production or as a feed for livestock.

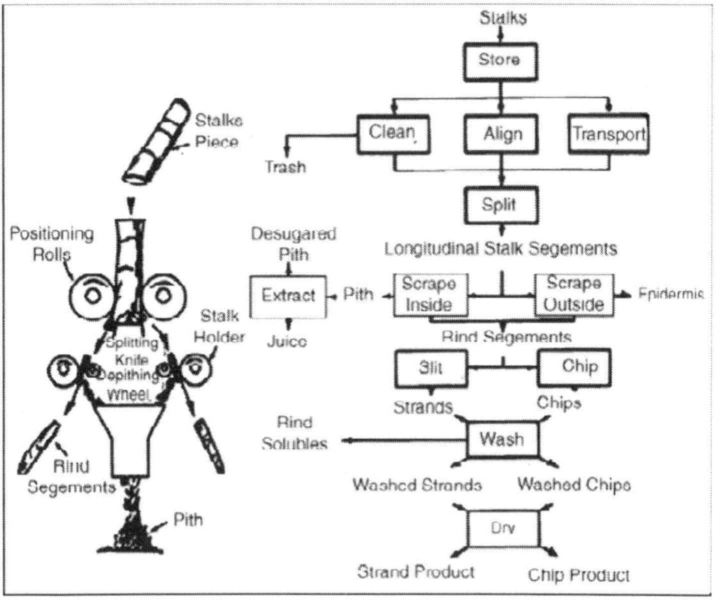

Fig. Simplified Diagram of the Tilby Separator Process for Derinding Sugarcane

In early experiments carried out by Donefer and his colleagues in Barbados, high rates of growth were observed in cattle fed sugarcane pith when it was supplemented with chopped cane tops and a concentrate c on taining an oilseed mea l and urea . Unfortun ately, theseprom is in g developments have not led to commercial application. The main limitation is the need for sophisticated and expen-sive machinery,necessitating a large- scale plant for economic viability. A factory producing compressed board would need a production capacity of about 30 tonnes ofboard per day, requiring the processing of 400 tonnes of sugarcane stalks. This would produce about 300 tonnes of pith per day-enough to feed15,000 head of cattle. While this size of unit is not unusual by feedlot standards in the USA, it is quite impractical for most developing countries,where the use of sugarcane as an animal feed has most potential.

A second constraint which emerged from research in Mexico (Preston et al. 1976) was the realisation that, as a feed for cattle, the pith was little betterthan the chopped whole sugarcane plant when both were adequately supplemented, even though sugar-cane pith contains a higher concentration of soluble sugars than chopped whole cane and thus has a potentially greater feed value. In fact when it was supplemented with ricepolishings (providing bypass protein, starch and oil) at low levels, the whole sugarcane supported higher growth rates than the derinded stalk. Only at the highest level of supplementation was there an indication that animal productivity was higher on a basal diet of pith. Nevertheless, differences were relatively small. These data indicate that it is the com-position of the supplement which determines animal productivity rather than minor variations in the ratio of sugar to fibre in the basal diet.

Meyreles et al. (1979) demonstrated that there is potential for exploiting the higher energy value of the pith by rational supplementation. In the absence of supplements other than urea and minerals, animals did not grow on a diet of sugar-cane pith. When green foliage of sweet potato was added to the diet, however, growth rate increased to about 500g/day and there was a significant re-sponse to high levels of urea (400 to 600g/ day). This can be interpreted as indicating an increase in micro-bial growth in response to an improved rumen microbial ecosystem, which. may be due to thesupply of both readily digestible,fibre and essential nutrients for micro-organisms (eg.amino acids, peptides, min-erals and, perhaps, vitamins): Supplementation with cottonseed meal gave a similar growth response, but apparently without stimulating rumen function be-cause feeding the higher level of urea had no effect on growth rate. The combination of these supplements was additive and feeding the high level of urea appeared to stimulaterumen microbial growth, as in-dicated by the 60% increase in growth rate.

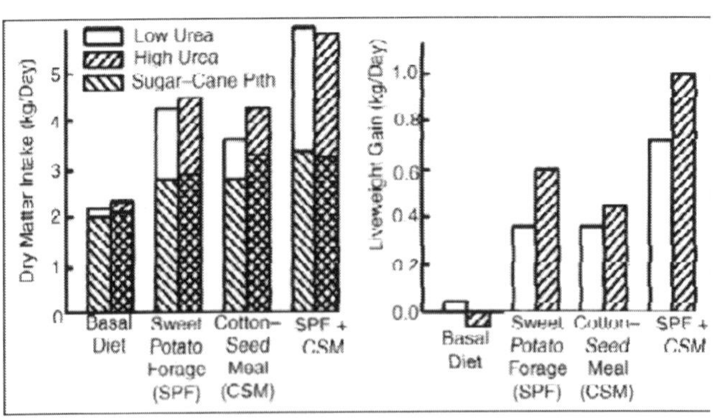

Fig. In a Cattle Fattening Diet Based on Ad Libitum Molasses, Sugarcane Tops and Wheat Bran (1 kg/d), Urea was a More Effective Source of Fer-mentable-N than Poultry Litter'. Poultry Litter Stimu-Lated Growth Rate when Added to the Diet (with Urea), which Wasappar'ently Adequate in Fermentable-N

Despite the possible advantages of using sugarcane pith rather than whole-cane, in some instances, the technology needed for chopping the wholesugarcane plant is simpler and much less sensitive to economies of scale than the derinding process. Chopping is therefore the preferred system for processing sugar-cane as a ruminant feed in developing countries.

FINAL MOLASSES

Alternative uses of Molasses as Animal Feed or for Power Ethanol

Molasses is the only "concentrated" source of fer-mentable carbohydrate that is widely available in the tropics and which is not a staple of the human diet. Its importance as a livestock feed is indicated by the ways in which it can be used:

- As a carrier for urea, minerals and other nutri-ents for improving the efficiency of utilisation of low-N diets (eg. crop residues, sugarcane and agro-industrial bypro ducts)
- A s a stra te gic dr o ug h t fe ed r es e rv e and a s th e b a s i s o f a supplement for routine feeding during the dry season.

Molasses is traditionally used to manufacture potable alcohol (eg. rum). H ow e ver, s i nc e , th e oil c ri s is en ormous e f for ts ha v e b e en ma d e to developproduction of industrial alcohol, especially as a sub-stitute for gasoline (power-alcohol). The lack of tech-nical knowledge of how to use molasses efficiently in livestock feeds has resulted in much molasses being discarded. Often molasses accumulates in the storage tanks at the sugar mills and is discarded into rivers. Its opportunity cost is thus very low, which has been a principal argument for its use in power-ethanol production. However, the recent developments in the use of molasses in livestock feeding have given profitable alternatives that are vastly superior economically to the production of power alcohol.

Table. Molasses may have a Higher Economic Value when Used as an Animal Feed than when it is Converted into Power Alcohol

	Opportunity	Assumption	cost $/t
Survival feeding of cattle in drought	540	1	
Substitute for cereals	150	2	
Fattening cattle	92	3	
Supplementing crop residues/dry pastures for:			
Milk production	440	4	
Prevent body weightloss	47	5	
Power alcohol	15	6	

- To keep cattle alive for 6 months about 360~:g mo-lasses/urea (1%) is required. The cost of a replace-ment animal is estimated at US$200. After sub-stituting the urea and distribution costs the value attributable to molasses is approximately $540 per tonne
- The value of molasses as a replacement for imported grain i .• estimated on the basis that it has a substitu-tion value of 75 % and assuming that imported grain is US$200/tonne (CIF)
- It is estimated that used as the basal diet (70% of the total feed) for fattening cattle 6kg molasses supplemented with 150g urea and 8kg fresh legume forage supports a daily liveweight gain of 750g. The urea cost.. $0.045, the forage US$0.46 and the liveweight gain is valued at US$0.87/kg
- The daily con.oumption of 500g of a molasses/urea block; (50% molaHes, 20% urea, 20% bran, 10% minerals) ha.. increased milk yield on a straw-based diet by lkg/day (milk is worth to the farmers US$0.20/kg
- A daily intake of 1. 5kg of 8 % urea/molasses mixture prevents a weight 10H in cattle of 200g/day and that this weight 10H costs US$0.50/kg 4kg molasses are fermented to produce 1 litre of an-hydrous alcohol. Costs that are not included are the fuel for distillation and the interest/ depreciation on the investment in a distillery. A barrel of oil (CIF) is assumed /,0 be worth $15.

There is an extremely strong argument for retain-ing molasses as a livestock feed in countries that suffer dry seasons and periodic droughts. The opportunity cost of using molasses (which may be the only feed available) in drought regions in the tropics is related to the value of a live animal and thesurvival of subsis-tence farmers who are dependent on draught animals.

Molasses is used widely in compounded feeds in the industrialised countries, where it is used to improve the palatability and binding properties of pelleted feed as well as reducing dustiness. Only low concentrations (5-10%) are needed for this purpose; higher levels make mixing and pelleting difficult. Because the molasses comprises such a small proportion of the diet, no nutrient imbalance is apparent. In tropical countries the quantities of molasses available are large relative to other potential feed ingredients.

In these situations molasses can be used in three ways:

- As a fermentable carbohydrate providing the ba-sis of the diet for ruminants
- As a palatable carrier for other essential nutri-ents (eg. urea and minerals) to supplement fi-brous diets and also a gelling agent in the production of nutrient blocks
- As a source of trace minerals and some macro-elements (eg. sulphur, calcium and potassium).

Many attempts have been made to incorporate mo-lasses into relatively inert materials such as bagasse and bagacillo (the residue after extraction of the coarse fibres for paper and particle board manufac-ture) but the final product is of poorer feeding value than the original molasses and is more expensive due to the high cost of power for mixing.

Supplementation

Research was conducted in Cuba in the late six-ties aimed at developing livestock feeding systems in which molasses was the principal ingredient. Atthe outset it was decided that molasses should be fed in its original liquid state in order to reduce processing costs and to facilitate transport and storage.

The successful development of the high-molasses fattening system for cattle (Preston et al. 1967) ex-emplifies the use of the basic principles ofruminant digestion and metabolism as outlined in earlier chap-ters. These include optimising rumen fermentation by supplying fermentable N (urea) andsome high-quality green forage and balancing the amount of protein (amino acids) to VFA by supplementing with a bypass protein source.

The original system used forages such as ele-phant grass, pangola grass and sugarcane tops as the roughage source. The availability of forage was restricted to 0.8kg dry matter/100 kg liveweight to encourage the animal to consume large amounts of molasses. The urea level was set at 2.5% of thefresh weight of the molasses to provide a ratio of fermentable N to carbohydrate close to the theoretical requirements of rumen micro-organisms.Sulphur supplementation was not required as sulphur diox-ide used in clarification of cane juice and the residual sulphur concentrated in the molasses.

In the widespread commercial application of the feeding system in Cuba, fish meal (Peruvian) was the bypass-protein supplement. The dramatic effect of this supplement in raising animal productivity on the molasses-based diet is shown in Figure.

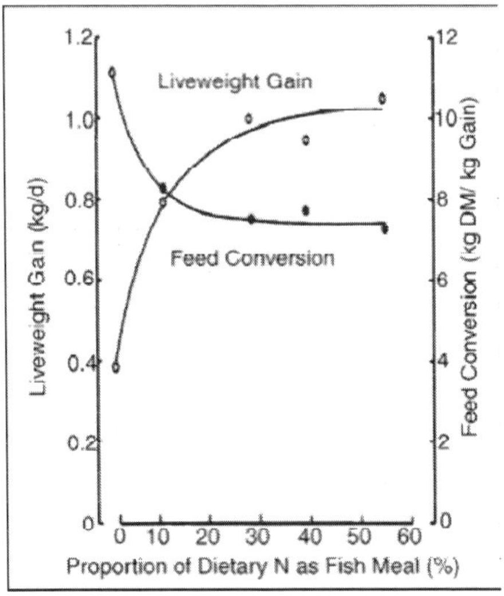

Fig. Addition of Fish Mealk to a Basal Diet of Ad Libitum Molasses-Urea and Restricted Forage Substantially Improved Growth Rate and Feed Conversion of Cattle in Cuba

Subsequent developments in the use of molasses-based diets have been directed to the use of:

- Protein-rich forages that supplied much, and some-times all, of the bypass protein as well as the roughage ; and
- Supplementation with poultry litter.

Table: Leucaena Foliage was as Effective as Groundnut Meal in Supporting High Growth Rates in Young Friesian Bulls Fed a Diet Based on Mo-lasses/urea. The Leucaena Replaced both the Forage (Native Grass) and the Groundnut Cake, and May thus have Served as a Combined Source of Protein and Roughage

	Leucaena (% LW/d)			Groundnut. Cake (g/d)		
LW gain (g/d)	790	740	847	597	742	
Feed conversion	9	12	10	10	10	(kg/kg gain)
Molasses (% of diet)	79	68	62	73	53	

Table: The Foliages of Cassava or Sweet Potato was used as the l'oughage Supplement in Amo-lasses-urea Diet for Fattening Bulls in the Presence or Absence of a Supplement of Soyabeanmeal. With Cas-sava Supplementation there were no Benefits from Ad-ditional Protein Indicating that it was a Better Source of Bypass Protein than Sweet Potato Foliage.

Forage	Sweet potato	foliage	Cassava	foliage
Soybean meal (g/d)	0	400	0	400
LW gain, g/d)	650	850	850	870
Molasses (kg/d)	5	6	6	6
Forage (kg/d)	13	13	10	10

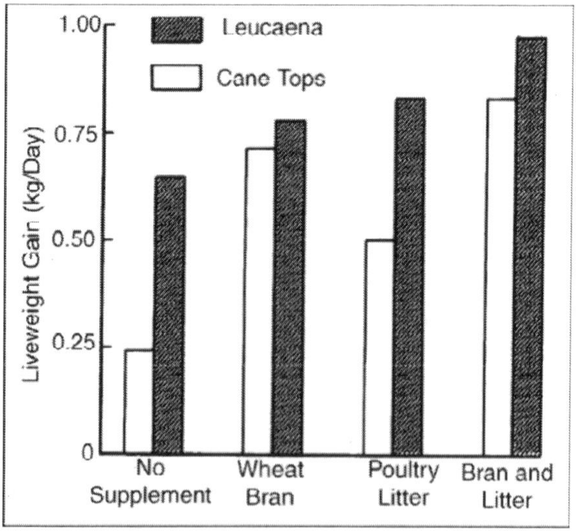

Fig. Contrasting Effects of Legume (Leucaena Foliage) and Grass (Sugarcane Tops) as the Roughage Supplement in a Diet Based on Ad Libitum Molasses/u1'ea with or with out Supplements of Wheat Bran

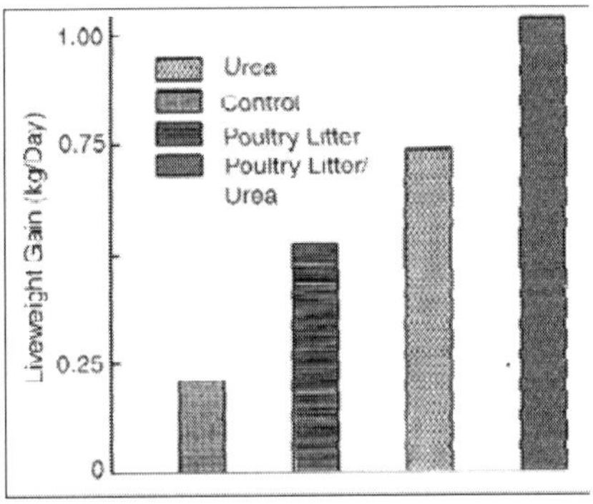

Fig. In a Cattle Fattening Diet Based on Ad Libitum Molasses, Sugarcane Tops and Wheat Bran (1 kg/d), Urea was a More Effective Source of Fer-mentable-N than Poultry Litter'.

Poultry litter stimu-lated growth rate when added to the diet (with Urea), which was appar'ently adequate in fermentable-N. There is evidence that, on molasses-based diets, poultry litter influences the pattern of VFA forma-tion, increasing the proportion of propionate and de-creasing the proportion of butyrate. This would partly explain the improved growth rates and feed conversions associated with the use of poul-try litter in molasses- based diets.

Commercial Feeding Systems Based on Molasses

The data m Tables summarise the results obtained when the molasses fattening programme was used commercially in large-scale feedlots and under conditions of restricted grazing on State farms in Cuba.

Table: Production Data from a Commercial Cattle Feedlot (10,000 Head) in Cuba During the Years when a Change was Made from an Ad Libitum Forage(Pan-gola Grass) Plus Concentrate Diet to one Based on Ad Libitummolasses/urea and Restricted Forage and Fish Meal. The High Losses Due Toemergency Slaughter and Mortality in the First Year of High Molasses Feeding were Due to Molasses Toxicity.

	Forage	Molasses-based	
	1969	1970	1971
Total LW gain (kg/d)	3,724	8,295	13,797
LW gain (g/head/d)	430	880	890
Feed conversion(kg DM/kg gain)	15	11	10
Deaths (%)	0.1	1.0	0.2
Emergency slaughter (%)	0.1	1.0	0.2

Table: Performance of Bulls Fattened Commer-cially in Cuba on Ad Libitummolasses/urea, Fish Meal and Restricted Grazing (3 hours/ day). Results are for 11 Unitseach with 400 Animals

	LW gain	Conversion (kg/kg LW gain)		
	(kg/d)	Molasses	Fish meal	Urea
Worst	0.74	14.7	0.54	0.47
Mean	0.83	9.1	0.45	0.29
Best	1.04	5.9	0.32	0.19

METABOLIC DISORDERS ON MOLASSES-FEEDING SYSTEMS

Three metabolic disorders may occur in cattle and sheep fed diets containing more than 50% molasses:

- Urea toxicity
- Molasses toxicity
- Bloat.

Urea Toxicity

This is discussed since, when high levels of molasses are used, urea is almost invariably included in the mo-lasses, and urea intakes may be as high as300g/ day (eg. in a 500kg dairy cow consuming 10kg of the molasses/urea mixture per day).

However, there is little risk of urea toxicity since, the sugars in molasses and ammonia from urea are quickly used in rumen microbial cell growth.Animals that have not previously consumed urea can be safely permitted free access to molasses containing up to 3% urea without fear of toxicity.Toxicity will only occur if the urea is not uniformly distributed, or mistakes have been made in formulation.

Molasses Toxicity

This is probably the most serious problem associated with molasses feeding. For example, in the first year following the introduction of the molasses/urea fattening system in Cuba, mortality and emergency slaughter rates in a 10,000 head feedlot increased from 0.1 % and 0.4% (when a forage- based diet was fed) to 1% and 3% respectively, when the diet was changed to high levels of molasses/ urea.

Ca ttle s uffering from molasses toxic ity salivate, sta nd apart in a "dejected" posture, usually with their heads lowered; and frequently are found "leaning" against the fence or feed trough. Invariably, eye-esight is affected and often the animal is blind. When disturbed they have an unsteady and uncoordinated gait and this led the "cowboys" in Cuba to refer to the affected animal as "borracho" (ie. drunk!!).

The nervous symptoms and blindness that were a feature of molasses toxicity indicated damage to the brain and it was subsequently shown by Verdura and Zamora (1970) that the clinical syndrome was indistinguishable f r om th a t of ce rebr o- cortic a l n e crosis (C C N) a lso k n own a s polioencephalomalacia (Edwin et al. 1979). The necrosis in the brain is readily seen and this allows rapid diagnosis.

The cause of the necrosis is likely to be a decrease in the energy supply to the brain because of either an absolute deficiency of alimentary thiamine, binding of thiamine analogues produced in the rumen and/or through the action of thiaminase in the rumen (Edwin et al. 1979), or a deficiency of glucose (Losada and Preston 1973). The evidence for the last explanation is that rumen propionate levels were extremely low (less than 10% molar) when the disease was induced in cattle deprived of forage (Losada and Preston 1973), and that daily oral administration of 400g of glycerol (a glucogenic substance frequently used in the treatment of ketosis) prevented clinical symptoms of the disease (Gaytan et al. 1977). In contrast, administering thiamine either intra-ruminally or intra-amuscularly did not prevent the disease (Losada et al, 1971).

Recent work by Rowe et al (1979a) introduced another element into the aetiology of molasses toxicity. They confirmed that removing the forage from a molasses-based diet induced symptoms of molasses toxicity, including brain necrosis, in which neither blood glucose level nor the rate of entry of glucose into blood was impaired. The most significant finding was that the absence of forage in the diet caused a rapid reduction in the rate of turnover of rumen fluid to the point of almost complete rumen stasis (from 1.5 to 0.5 volumes/day). Rowe et al. (1979a) hypothesized that, under these conditions, the supply of both amino acids and B-vitamins to the animal would be drastically reduced with a consequent decrease in the availability of thiamine for brain metabolism. Another explanation is that low rates of turnover of ru-men fluid might. encourage the proliferation of slow--growing bacteria that produce thiaminase.

It is apparent that a number of factors interact in the development of molasses toxicity. The basic defect common to all situations isthe disruption of energy supply to the brain. This could be brought about by inadequacies in the ab-solute supply of blood glucose or thiamine (requiredfor brain metabolism of glucose) or both, as well as by the action of thiaminases present in the feed or produced by the rumen microorganisms.

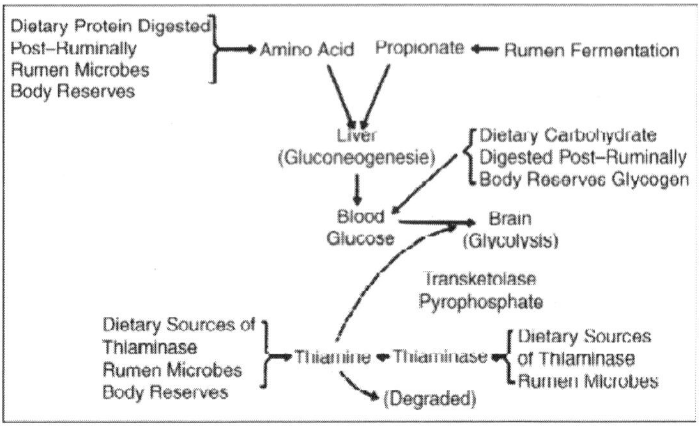

Fig. A Summary of some Factors that may Affect the Supply of Blood Glucose and Itsutilisation by the Brain

The theory that glucose deficiency is the main cause of molasses toxicity seems the most plausi-ble. The availability of glucose to cattle on molasses- based diets is always precariously balanced because of the low-propionate/ high-butyrate fermentation of molasses, and t.he economic pressure to mini mise true-protein inputs. Molasses toxicity does not ap-parently occur in sheep. Sheep fed such diets have higher propionatefermentation patterns than c.attle which lends support to the concept of glucose being primarily involved in the syndrome. Removal of for-age from amolasses-based diet givell to sheep does not result in the same rumen stasis as in cattle.

Two processes may separately, or in combination, lead to excessive use of glucose and a lowering of blood glucose concentrations, which could lead to brain damage.

These are:

- Periods of exercise/excitement, which illcrease the utilisation of glucose, especially by muscle
- Consumption of an excessively large amount of molasses at one time, leading to a requirement for glucose to clear the increased amounts of ac-etate absorbed. Molasses ferments very rapidly in the rumen and animals normally eat "little and often" throughout the day. Disturbance in this pattern, brought about by poor manage- ment or inadequate trough space and therefore "dominance" by some animals, could be the trig-gering mechanism for the over- eating.

This thesis is supported by observations of periods of low blood glucose levels in cattle during experi-ments which involved intensive blood sampling (R A Leng and M H Ferreiro, unpublished; also Gaytan et al. 1977).

When feeding high-molasses diets to cattle it is usual to restrict the supply of forage, either to in-crease intake of molasses or because of the greater cost of forage compared to molasses.

Inadequacies in either the quantity or quality of forage appear to be the main causes of molasses toxi-city. Thus, the incidence of molasses toxicity was less when wheat or barley straw, rather than sorghum for-age or maize silage, were used as the forage sources in molasses-based diets. Furthermore, there have been no reports of toxicity when high-protein forages (eg. leucaena and cassava alld sweet-potato tops) have been used. Equally, feeding nutritious and palatable forage appears to be the best cure for affected animals (T R Preston, unpublished data).

Recent developments in the molasses-feeding sys-tem have emphasised the technical and economic ad-vantages of feeding high-protein forages, especially from leguminous plants such as leucaena, as a com-bined source of both roughage and bypass protein. Including such forages in the diet is also likely toof-fer the most cost-effective means of avoiding molasses toxicity.

Bloat

Bloat is the retention in the rumen of gas, either free or trapped in foam, and occurs in cattle on almost all feeding systems. It is more prevalent withsome diets (eg. grazed legumes that contain high levels of soluble foaming agents such as saponins and proteins). It is not a partic.ularly serious problem with molasses feeding. However, relatively high incidence of bloat (up to 5%) has been reported where access to the molasses is occasionally limited(eg. grazing cattle allowed access to molasses/urea only once or twice daily) and where cattle have eaten relatively large quantities of molasses/ ureaover short periods.

A hypothesis for the aetiology of the syndrome is as follows. In animals fed molasses based diets, an organism grows in the rumen which produces consid-erable mucilaginous secretions and which can reach high population densities. When the animals consume molasses quickly, the rate of fermentation in the rumen is high (producing considerable quantities of carbon dioxide and methane), the pH falls and the bicar bonate in the rumen fluid is converted to CO_2. The very rapid production of gas in the presence of the mucilagenous organism forms a foamy bloat. A similar train of events is reported to cause feedlot bloat on grain-based diets. An optimum level of rumen ammonia accelerates fermen-tation and therefore bloat may be controlled by using a source of NPN that releases ammonia more slowly than urea, eg:chicken litter.

SUGAR CANE JUICE

The extremely low fermentability of sugarcane fibre and the negative effect this had on vol-untary intake of the overall diet was the reason for developing methods for fractionating the crop so that sugar and fibre could be treated as separate entities.

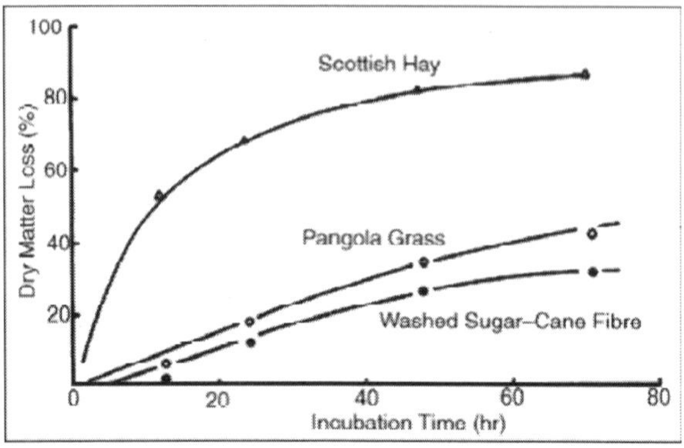

Fig. Rate of Loss of Dry Matter from Ny-lon Bags Containing Washed Sugarcane Fibl'e, Pangola Grass or a Sample of Hay Grown in Scotland, Suspendedin the Rumen of a Steer Fed a Sugarcane-based Diet

The new approach to fractionation of sugarcane for livestock feeding and fuel production was first devel-oped in Mexico (Preston 1980b). The aim was to achieve only partial extraction of the juice using sim-ple low-cost cane crus hers, of the kind developed for "panda " an d "gur" production. Thisreduced the investment in machinery and the energy cost of milling (stalks are passed only once through the crusher).

The justification for this system was that the juice is a suitable carbohydrate resource for both ruminant and non-ruminant livestock. The residual sugar-rich fibre can be gasified to make "producer gas" to sub- stitute for gasoline and diesel oil (T R Preston and A Lindgren, unpublished data). Alternatively it can be used as a component of the diet of draught animals or small ruminants that can readily select the pith from the rind.

Supplementation

The fermentable carbohydrates in sugarcane juice and molasses are sucrose, glucose and fructose. Molasses is the concentrated soluble residues after extraction of the sucrose from cane juice. It is therefore richer in minerals, organic acids and other soluble plant components than cane juice.

Diets based on cane juice support much higher levels of an imal productivity than those based on mo-lasses. The rates of growth and feed conversion efficiency recorded in studies in Mexico are comparable to those recorded in intensive grain-based systems.

Table: Holstein/Zebu Bulls had Higher g1'Owth Rates and Better Feed Conversion on a Basal Diet of Fresh Sugarcane Juice than on Molasses. The Differ-ence was Especially Marked in the Absence of the By-pass Protein Supplement (Sunflower Seed Meal SSM).

Basal diet SSM supplement(kg/d)		LW gain (g/d)	Intake of molasses/cane juice (kg/d)	Feed conversion (kgDM/kg gain)
Molasses	0.0	250	4	26
	1.0	550	4	12
Cane juice	0.0	800	23	7.4
	1.0	1320	32	6.4

The potential for use of cane juice is considerable because microbial growth in the rumen on this feed appears to be highly efficient; Zebu cattle grew at800g per day on minimal protein intake (5% protein in the diet dry ma tter supplied from leucae na) with the fermentable N provided by aqueousammonia. This is in contrast to the average grain-based feeding system, in which at least 12% of the diet dry matter is protein. It appears that on a canejuice diet, the end-products of rumen fermentation are better bal-anced and therefore support much higher levels of an-imal productivity than sugarcanepith. This indicates that microbial growth in the rumen was extremely efficient since, it must have supplied an almost ideal protein/energy ratio for production without supple-mentation.

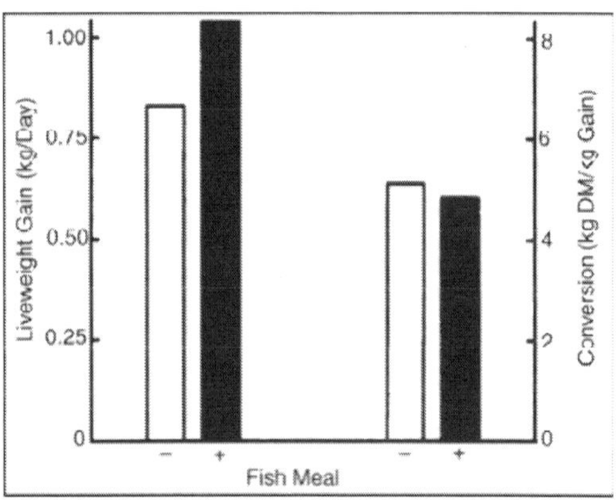

Fig. Zebu Steers had High Growth Rates and Efficient Feed Conversion Onsugar Cane Juice Supple-mented with Ammonia and Leucaena Foliage. Addi-tionalbypass Protein from Fish Meal Increased Growth Rate Slightly

Two observations on feeding cane juice to cattle are pertinent to the underlying thesis of this book:

1. Propionate comprised more than 20% of the total VFAs in the rumen, giving a high glucogenic energy ratio (G/E) in the animals. In this respect the juice was dearly superior to molasses
2. The population of protozoa in the rumen of an-imals fed cane juice was lower than that in the rumen of molasses-fed animals (ie. 10,000/ml versus

1,000,000/ml of rumen fluid), again suggesting efficient microbial protein production (D Harrison , personal communication; Bird and Leng1978).

The importance of these factors as determinants of efficient ruminant productivity are discussed.

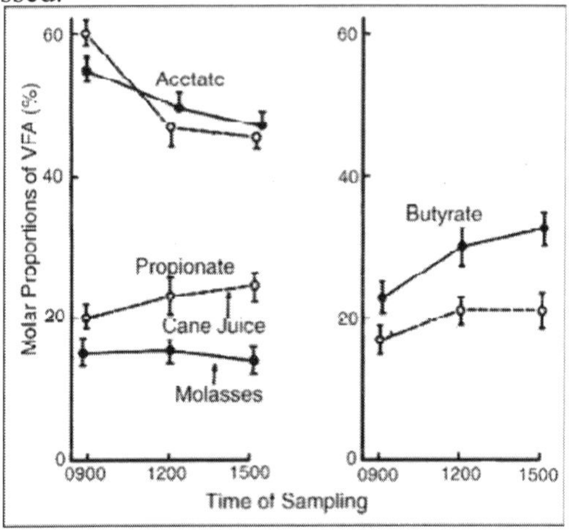

Fig. There were Higher Molar Proportion of Propionic Acid and Lowerproportion of Butyric Acid in Rumen Fluid of Cattle Fed Sugarcane Juice than with Cattle Fed Molasses

RESIDUAL PRESSED CANE STALK

The maximum economic benefit from the use of sug-arcane juice for livestoc.k feeding will only be realised iftechnologies are developed to use theresidual pressed stalk.

The options being pursued for the use of this material include:

- Productiojn of charcoal
- Conversion to producer gas (a mixture of carbon monoxide, hydrogen and nitrogen) as a substi-tute for diesel fuel
- Feeding to draught animals
- Feeding to small ruminants (sheep and goats), allowing a high degree of selection (eg. by providing twice the normal dry-matter in take).

Making charcoal from pressed cane stalk is simple but requires briquetting to produce a readily saleable product (Ffoulkes et al. 1980).

To be able to use pressed cane stalk for producer-gas production, it must first be sun dried to 20% moisture content and chipped into particles of 10 to 20mm in length. A gasifier has been designed that uses dried sugarcane chips (A Lindgren, personal communication).

A promising development is the potential of using the pressed (extracted) cane stalk as a basal diet for small rum in ants. W h en it is a vailable immediatelyfollowing juice extraction, goats eat it avidly, selecting the sugar- rich pith and rejecting the lignified rind. In studies in Colombia, goats on a mixed diet ofpressed stalk and fresh gliricidia foliage selected, and apparently preferred, the pith to the green foliage (Preston 1987).

SISAL BAGASSE AND PULP

INTRODUCTION

Sisal (Agave fourcroydes) is a source of fibre for twine, ropes and carpets. The entire leaf blade is harvested and transported to the factory, and there are thus no residues in the field. However, the bagasse that remains after th e fibr es ar e extr a cted is poten ti al ly use f ul. Th e b aga sse le ft a ft er traditionalprocessing contains both long and short fibres. More advanced factories extract a greater proportion of the fibres and some of the saponins, which areused as raw ma-terial for the manufacture of steroid hormones. The byproduct of this modified process is referred to as "pulp" to differentiate it frombagasse (Riley 1984).

Data on the chemical composition of both these byproducts are given in Table. Because of the high content of soluble sugars and the high moisture content, the pulp or bagasse begins to ferment imme-diately after fibre extraction and most of the soluble sugars are converted to lactic acid.

Table. The Chemical Composition of Sisal Bagasse and Pulp

	Sisal pulp		Sisal bagasse		
	Fresh	Ensiled	Fresh	Ensiled	Fresh
Proximate analysis (% DM)					
Crude fibre	29	30	28	29	23
Crude protein	5.3	4.9	5.3	5.5	6.3
Ether extract	3.1	3.0	3.7	3.4	3.0
Ash	13	13	15	15	12
Water-soluble CHO	21	0	27	0	
Macl'O-minerals (g/kg DM)					
Calcium	51	53	47	49	35
Phosphorus	1.1	1.0	1.0	1.2	1.5
Magnesium	8.0	8.0	9.0	10.0	8
Trace minerals (mg/kg DM)					
Zinc	1.2	1.2	1.5	1.4	
Copper	0.5	0.5	1.0	0.8	
Cobalt	ND	ND	ND	ND	
Manganese	1.0	1.1	1.0	0.8	
Iron	2.0	1.8	2.3	2.2	
Iron	2.0	1.8	2.3	2.2	

Organic acids (% DM)					
Lactic	1.6	16.5	1.0	18.1	
Citric	1.0	0.5	1.2	0.8	
Oxali,	5.5	5.3	5.2	5.4	
Ph	4.0	3.9	3.9	3.8	

The constraints associated with the use of the pulp/bagasse as a feed for ruminants relate partly to the content of organic acids, principally lactic andoxalic acids. Phosphorus and some of the trace ele-ments are also severely deficient, and thus the min-erai content of pulp or bagasse is highlyimbalanced relative to the needs of productive animals.

Cattle fed only on ensiled sisal pulp for a long pe-riod developed acidosis and barely maintained body-weight, despite the reasonably high digestibility(50-60%) of the feed. Attempts to correct only the acidosis and the mineral imbalance did not increase animal productivity (Naseeven and Harrison 1981; Belmar and Riley 1984).

Research in Mexico on the use of sisal pulp as a feed for ruminants (Rodriguez 1983) is a further example of the application of the principles developed in this book, namely that attention should first be given to optimising the rumen ecosystem and then to correcting the protein/energy ratio with bypass protein. Only after these two steps have been taken are responses to supplementary minerals observed.

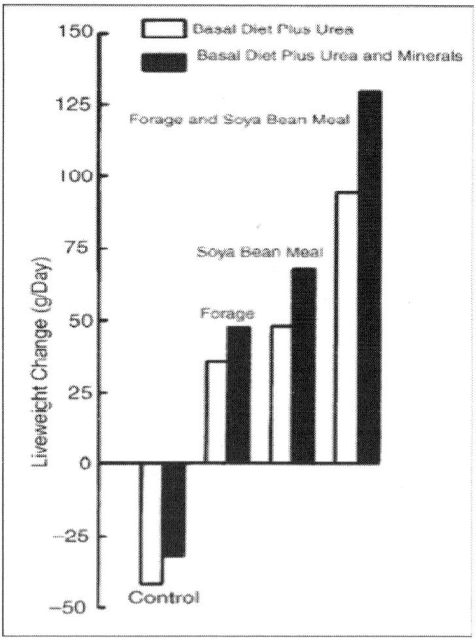

Fig. Sisal Pulp is Relatively Digestible (55-60%) but Low in Fennentable-N, Protein and Minerals.

BANANAS AND PLANTAINS

Bananas and plantains provide voluminous amounts of crop residues which are normally incorporated into the soil for mulching and to maintain soil fertility. When bananas are grown for export, up to 20% of the fruit may be rejected, and these are often discarded.

There are few nutritional constraints to the use of banana fruit as feed for ruminants. However, it is important to feed the material while it is green andthe carbohydrate is still in the form of starch. The fruit can be fed as a complete bunch (or "hand") and there is no need for any processing. The fruit islow in nitrogen and the skin is rich in tannins. The greatest nutrient deficiency of bananas as a feed for ruminants is fermentable nitrogen and dietscontain-ing bananas must be supplemented with urea.

Where molasses is freely available, the most con-venient procedure for supplementing cattle with urea is to allow the cattle free access to aurea/ molasses mixture (10% urea). Cattle will restrict their intake of this supplement to about 2kg/day, which provides enough urea (200g/day) to ensure an adequate level of ammonia in the rumen for efficient fermentation (McEvoy and Preston 1976). If a high-protein forage is fed with the diet,either as a legume component of a pasture in a restricted-grazing system or as a protein- rich foliage (eg. leucaena or gliricidia), there appears to be no advantage from supplementing with a bypass-protein meal. Results from feeding reject bananas to cattle in Colombia are summarised in Table.

Table: Growths Rates of Cattle Fattened Under Commercial Feedlot Conditions in Colombia on Reject Bananas (40% of Diet DM), Chopped Elephant Grass (42%) and Kudzu Legume Forage (18%) with Different Supplements.

Protein suppl.	Initial LW feed	Daily gain	Days on	Comparisons	(kg)	(kg/d)
No supp!.	330	0.83	90	Female adult		
	325	0.64	90	Male adult		
	334	0.80	90	Female old		
	325	0.88	90	Female young		
	240	0.91	100	Male young		
	350	0.61	100	Male old		
1kg/d CSC	381	1.15	60	+ sulphur		
	391	0.97	60	– sulphur		
	332	0.98	60	Ureal		
				Jllolasses		
	322	0.97	60	Aqueous urea		
2kg/d CSC		1.60		Chianina×Zebu		
		1.14		Zebu crosses		
		1.25		Zebu bulls		
		1.10		Zebu steers		
		1.66		Chianina bulls		
		1.50		Chianina steers		

Banana Foliage

The dry matter in the leaves and pseudostem of the banana plant is relatively digestible (65 and 75%, respectively; Ffoulkes and Preston 1978a).Despite this, these feeds barely support maintenance if given alone to ruminants. However, when ruminants are fed banana pseudostem they respond to supplementation with urea, highly digestible green forage and bypass nutrients. The data in Figure show that cat-tle increased their voluntary feed intake significantly when the chopped leaves and pseudostems were sup-plemented with sweet-potato foliage or with wheat bran. The highest feedintake was observed when both sweet- potato foliage and wheat bran were given together.

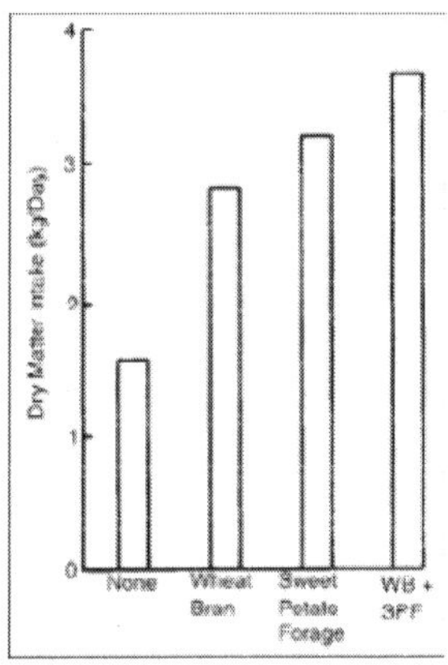

Fig. Supplements of Wheat Bran (1kg/d) and/or Sweet Potato Tops (5% Bodyweight, Fresh Basis) Increased the Voluntary Feed Intake of Cattle given a Basal Diet of Chopped Banana Pseudostcm, Urea and Minerals

Feeding systems based on banana foliage have been developed in the Seychelle Islands. Ac-ceptable levels of growth were obtained incattle fed the banana forage, urea and leucaena, and there was an apparent response to additional supplementation with banana fruit.

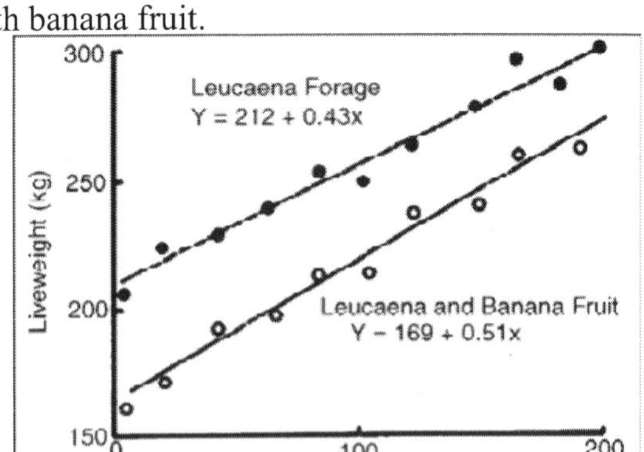

Fig. Growth Performance of Crossbred (Jerrsey/Creole) Cattle in Seychelles Fed a Basal Diet Based on the Whole Banana Plant (Pseudostem and Leaves Plus Urea and Minerals) and Supplements of Fresh Leuucaena Foliage (3% Bodyweight/d) Alone or with Some Reject Banana Fruit (3kg/d) (M Delpeche and T R Preston, unpublished data).

Sugarcane produces enormous amounts of biomass. Picture shows sugarcane being chopped for feeding trials with cattle in Mexico (Alvarez F)

Crop residues are the major feed biomass for cattle 111 developing countries. ❒

3

Principles of Plant Genetics and Breeding

WHAT IS A GENETIC PROGRAMME

Much of the theory behind genetic programming is the same as that behind genetic algorithms. The same Darwinian concept of survival-of-the- fittest applies, through genetic operators, but with a twist.

The structures that are manipulated are quite different from the coded strings of genetic algorithms. Figure depicts how one might visualize a simple genetic programme.

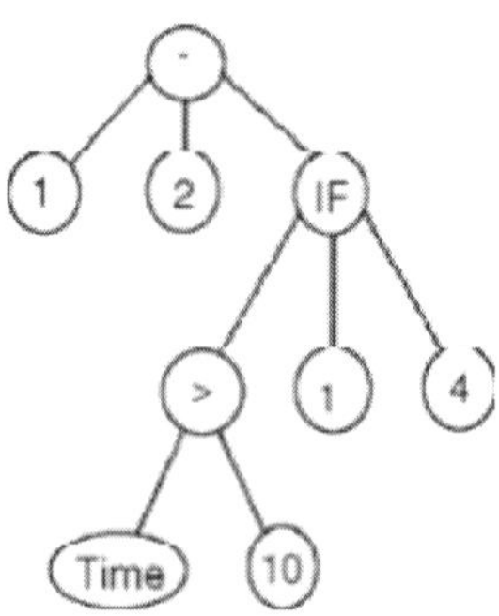

Fig. A Simple Tree Structure.

The equivalent LISP expression for this function is (+ 1 2 (IF (> TIME 10) 3 4)).

A few notes:

- There now exists a clean hierarchical structure, instead of a flat, one-dimensional string;
- The structure is made up of simple functions, that can be easily encoded using a high-level language;
- Tree manipulation routines exist in several high-level languages.

GENERATING A POPULATION OF PROGRAMMES

Genetic algorithms require the generation of random, encoded strings for searching the solution-space. Under the genetic programing approach, a population of random trees representing programmes is constructed. The genetic operators (crossover, reproduction, etc.) are performed on these trees, in a manner that will be described in the next subsection.

In order to create these individuals, two distinct sets are defined: the terminal set T, and the function set F. For the case of depicted in Figure, {1,2,3,4,10, TIMEoT}, and {+,>,IF}oF. Notice that the terminal set can include variables, as well as constants. All the functions and terminals must be compatible (i.e., pass information between each other).

In general, to make a random tree, then, is to pick randomly from T U F, until all the branches end in terminals. Of course, one can skew the probabilities to obtain trees of certain sizes, as well as place absolute limits on the depth of tree.

To generate a population of programmes, just perform this initialization procedure as many times as needed. The next issue to tackle is how to reformulate the operators of genetic algorithms into this new framework.

GENETIC ALGORITHMS

Genetic programming is a branch of genetic algorithms. The main difference between genetic programming and genetic algorithms is the representation of the solution.

Genetic programming creates computer programmes in the lisp or scheme computer languages as the solution. Genetic algorithms create a string of numbers that represent the solution.

Genetic programming, uses four steps to solve problems:

1. Generate an initial population of random compositions of the functions and terminals of the problem (computer programmes).
2. Execute each programme in the population and assign it a fitness value according to how well it solves the problem.
3. Create a new population of computer programmes.
   - Copy the best existing programmes
   - Create new computer programmes by mutation.
   - C re at e n e w c o m puter progra mm es by crosso v e r(s ex u al reproduction).

4. The best computer programme that appeared in any generation, the best-so-far solution, is designated as the result of genetic programming.

FITNESS FUNCTION

The most difficult and most important concept of genetic programming is the fitness function. The fitness function determines how well a programme is able to solve the problem. It varies greatly from one type of programme to the next.

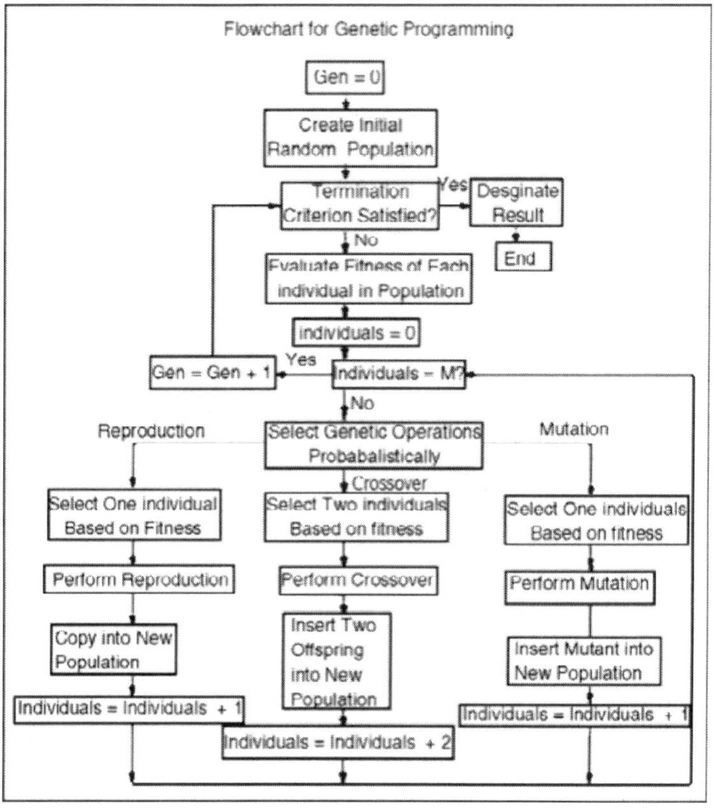

Fig. Genetic Programming Flowchart.

For example, if one were to create a genetic programme to set the time of a clock, the fitness function would simply be the amount of time that the clock is wrong. Unfortunately, few problems have such an easy fitness function; most cases require a slight modification of the problem in order to find the fitness.

GUN FIRING PROGRAMME

A more complicated example consists of training a genetic programme to fire a gun to hit a moving target. The fitness function is the distance that the bullet is off from the target. The programme has to learn to take into account a number of variables, such as wind velocity, type of gun used, distance to the target, height of the target, velocity and acceleration of the target. This problem represents the type of problem for which genetic programmes are best. It is a simple fitness function with a large number of variables.

WATER SPRINKLER SYSTEM

Consider a programme to control the flow of water through a system of water sprinklers. The fitness function is the correct amount of water evenly dis tributed over the s urface. Unfortunately, th ere is no one variable encompassing this measurement. Thus, the problem must be modified to find a numerical fitness. One possible solution is placing water-collecting measuring devices at certain intervals on the surface.

The fitness could then be the standard deviation in water level from all the measuring devices. Another possible fitness measure could be the difference between the lowest measured water level and the ideal amount of water; however, this number would not account in any way the water marks at other measuring devices, which may not be at the ideal mark.

MAZE SOLVING PROGRAMME

If one were to create a programme to find the solution to a maze, first, the programme would have to be trained with several known mazes. The ideal solution from the start to finish of the maze would be described by a path of dots. The fitness in this case would be the number of dots the program me is able to find. In orde r to prevent the programme from wandering around the maze too long, a time limit is implemented along with the fitness function.

FUNCTIONS AND TERMINALS

The terminal and function sets are also important components of genetic programming. The terminal and function sets are the alphabet of the programmes to be made. The terminal set consists of the variables and constants of the programmes. In the maze example, the terminal set would contain three commands: forward, right and left.

The function set consists of the functions of the programme. In the maze example the function set would contain: If “dot” then do x else do y. In the gun firing programme is the terminal set would be composed of the different variables of the problem.

Some of these variables could be the velocities and accelerations of the gun, the bullet and target. The functions are several mathematical functions, such as addition, subtraction, division, multiplication and other more complex functions.

CROSSOVER OPERATION

Two primary operations exist for modifying structures in genetic programming. The most important one is the crossover operation. In the crossover operation, two solutions are sexually combined to form two new solutions or offspring. The parents are chosen from the population by a function of the fitness of the solutions. Three methods exist for selecting the solutions for the crossover operation. The first method uses probability based on the fitness of the solution.

Another method for selecting the solution to be copied is tournament selection. Typically the genetic programme chooses two solutions random. The solution with the higher fitness will win. This method simulates biological mating patterns in which, two members of the same sex compete to mate with a third one of a different sex. Finally, the third method is done by rank.

In rank selection, selection is based on the rank, (not the numerical value) of the fitness values of the solutions of the population. The creation of the offsprings from the crossover operation is accomplish by deleting the crossover fragment of the first parent and then inserting the crossover fragment of the second parent.

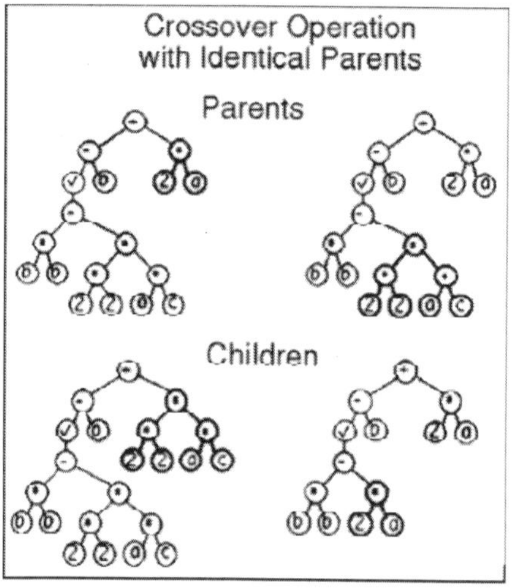

Fig. One of the Main Advantages of Genetic Programming Over Genetic Algorithms.

In genetic programming identical parents can yield different offspring, while in genetic algorithms identical parents would yield identicaloffspring. The bold selections indicate the subtrees to be swapped. An important improvement that genetic programming displays over genetic algorithms is its ability to create two new solutions from the same solution. In the Figure the same parent is used twice to create two new children.

MUTATION

Mutation is another important feature of genetic programming. Two types of mutations are possible. In the first kind a function can only replace a function or a terminal can only replace a terminal. In the second kind an entire subtree can replace another subtree. Figure explains the concept of mutation:

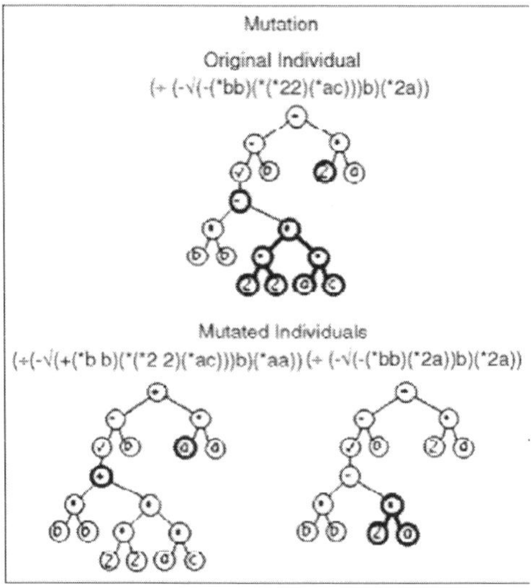

Fig. Two Different Types of Mutations.

The top parse tree is the original agent. The mutation of a single terminal for another single terminal (a). It also illustrates a mutation of a single function (–) for another single function (+). The parse tree on the bottom right illustrates a the replacement of a subtree by another subtree.

Genetic programming is much more powerful than genetic algorithms. The output of the s is a quantity, while the output of the genetic programming is a another computer programme. In essence, this is the beginning of computer programmes that programme themselves. Genetic programming works best for several types of problems. The first type is where there is no ideal solution, (for example, a programme that drives a car). There is no one solution to driving a car. Some solutions drive safely at the expense of time, while others drive fast at a high safety risk. Therefore, driving a car consists of making compromises of speed versus safety, as well as many other variables. In this case genetic programming will find a solution that attempts to compromise and be the most efficient solution from a large list of variables. Furthermore, genetic programming is useful in finding solutions where the variables are constantly changing. In the previous car example, the programme will find one solution for a smooth concrete highway, while it will find a totally different solution for a rough unpaved road.

GENETIC ALGORITHMS - NATURE'S WAY

In the vast eons of time the processes of evolution have created a common basis for all life on Earth - that of DNA or its simpler variant RNA. Both systems share a method of coding with that greatest of human inventions, the computer, yet the mode of operation of the two processes could hardly be more different.

Approaches to Achievement

In our use of computers we first decide an overall goal, we then break down the problem into manageable chunks and write step by step procedures (algorithms) by which to accomplish them.

This mirrors our normal approach to life, in which we decide what we wish to do and plan out a way of achieving our ambition, taking into account the resources and time needed. If the plan doesn't work we modify it and try again-aserial, one at a time approach to achievement.

Nature has a different way, brute force and ignorance. If there are a hundred possible options available, well try them all — the first to get there wins. All options are tried simultaneously — a parallel approach to achievement. But where is there? It just doesn't matter! At first sight, to human eyes, this is a strange concept — how can you get anywhere useful without knowing where it is?

TEACHING VERSUS EXPLORING

"Ignorance is bliss" goes the saying and we humans certainly start from a state of ignorance. Each of us can only plan for what we already know, the constraints of our limited knowledge restrict what we can imagine - we often cannot see beyond our assumptions.

To overcome these limitations we must rely on being taught new ideas by someone whom we trust, or by stumbling onto new discoveries by accident

- in both ways we expand our mental horizons and thus see new possibilities that were previously unthinkable. By adopting nature's method and trying every possible option we can arrive at every destination simultaneously. Thus we can enjoy every possible experience. Wow, what a lifestyle ! Is there a snag ? Of course.

We consume a hundred times as many resources for every hundred possible options - and the options increase geometrically (like sand grains doubling for every square on a chessboard), the universe just isn't big e n ough . Note however, that the conformist social equivalent of everybody doing exactly the same thing (all making the same mistake...), uses just as many resources, yet achieves nothing better - despite the considerableduplication waste involved, our planet isn't big enough for this either !

EVALUATING OPTIONS

But nature has the solution already, tree pruning. Imagine we have two options, each of those has two more and so on, drawing those out we get a picture like a branching tree.

Suppose we impose a test on each option, if the test fails then that option (branch) is deleted (in this illustration we eliminate any mouse genes that fail to find the central food). Once we do this we have pruned the tree of all failed branches and those left passed all our tests (this is similar to how Chess computer programmes work). But which test does nature use ? A very simple one - survival itself. Any organisms unable to survive and reproduce will die out, their genes pruned from the available options.

Can computers operate as nature does ? Yes, this is the general field known as Evolutionary Computation, of which the Genetic Algorithm is the basic technique and can be further expanded to remove the need for an underlying programme.

THE GENETIC ALGORITHM

How does it work? Well, we create a population of individuals, each represented by a chromosome (a collection of genes or characteristics) appropriate to the problem we are trying to investigate. Genes are usually implemented by a series of computer bits, enough to cover all available alternative values. The value of each gene is initially assigned randomly, within the available parameters.

The population is then evaluated to determine how well each individual does the required task. The best members of the population (typically the top 10per cent) then become the parents for the next generation. The genes of the offspring are created by selecting two parents at random and combining part of the chromosome of each, so parent gene combinations ABC mated with DEF may become two offspring ABF and DEC. This mimics the crossover or recombination of sexual reproduction in nature and is repeated for further pairs until a full second population is created. Random changes to individual bits are then made mimicking the natural role of mutation, at some desired rate (say 1 in 100 bits changed).

The resultant population is then evaluated as before, and the process is repeated as many times as desired until the required performance level has been achieved (or no further improvement seems possible). This technique can rapidly covers the space of all possible options and converge on a solution that is beyond the ability of all but the best human programmers, in areas where no conventional solution techniques exist.

COEVOLUTIONARY PROGRAMMING

In the Genetic Programming extension, the genes are the valid statements within a computer language and the chromosome in this case is itself the programme. Additional operations are allowed to join or split programme segments, allowing variable length programmes to evolve. By GA techniques (excluding mutation) we can thus allow a programme to create itself to solve a particular problem, discarding all the failed programmes. This has proved a very powerful way of generating solutions to very difficult problems, but with the drawback that the resultant code is all but incomprehensible to humans !

There is a difficulty here. If our test is too specific, then the programme might be perfect for that one task, but fail completely for very similar tasks. How can we overcome this ? Again by copying nature. Suppose we wish to create a programme to sort a series of numbers. If we have a fixed test sequence of numbers we encounter the problem mentioned, so let us have many sets of numbers, as many as we have candidate programmes.

Programmes can then encounter different tests at each stage and only those capable of a general solution will survive. But we can do better still, let us evolve the tests also by the same process, the test survivors will the ones that defeat a programme. Using this coevolution strategy, the tests get harder as the programmes get better and this bootstrapping feedback technique gives a considerable improvement in speed and efficiency. If rather than programmes and tests we have interacting agents we enter the realm of Artificial Life.

Evolutionary Thinking

Can we use these techniques in human thought processes ? Perhaps we already do. Whenever we think of a task do we not automatically generate many possible solutions and then evaluate them (subconsciously in most cases) against our goal, discarding all those that fail our criteria ? We then refine the survivors (combining the best parts of each) and so on until we have an optimum solution. Nature has again beaten us to it...

Socially a similar process is common. Brainstorming is a popular management technique for generating diverse pseudo-random solutions to a problem, equivalent to nature's mix and match juggling of genes perhaps. The group then evaluates (tests) the options and prunes those failing some criteria, the winners then propagate to the next stage and perhaps ultimately into production. Along with top down (analysis) and bottom up (synthesis) strategies - both goal directed, we also operate a parallel evolution strategy, companies (alternative chromosomes) compete to grow, ideas (memes - cultural genes) compete for attention. Evolution is embedded in our behaviour.

ART AND INTERACTIONS

The technique of Genetic Algorithms is widely applicable, even in art. If we code genes to refer to some feature of an image or music we can evolve extraordinary organically based artforms or music. Any system of building blocks can be readily adapted to these procedures and in this way transcend automatically the imagination of its creator. By constraining the system by the desired criteria (tests) we can allow design processes to evolve as a background task - resulting in quite unexpected innovations.

Nature's way. We must however be clear that we are not fully imitating nature he re. Our artificial genes (called G-Type s) have well defined correspondences with function or form (called P-Types). We also have only one chromosome of very limited length in our genome, whereas natural genotypes contain many chromosomes, each with thousands of genes.In nature genes translate to body form (phenotype or morphology) via a very complex interactive process, the form is itself an emergent phenomena dependent on the interaction of genes with themselves, with the cell chemistry and with the surrounding environment.

We cannot predict this process adequately without understanding the nature of emergence. Genetic engineering, whilst highly useful, is a very dangerous process. The side effects of any changes to a such a complex and highly non-linear process as life are, to say the least, unpredictable.

GENETIC PROGRAMMING OPERATORS

The four general genetic operators are: crossover, reproduction, mutation, and inversion. The following overview describes each in brief.

- Crossover: Koza considers crossover, along with reproduction, to be the two foremost genetic operations. It is mainly responsible for the genetic diversity in the population of programmes. Similar to its performance under genetic algorithms, crossover operates on two programmes (a binary operator), and produces two child programmes. Two random nodes are selected from within each programme and then the resultant "sub-trees" are swapped, generating two new programmes. These new programmes become part of the next generation of programmes to be evaluated.

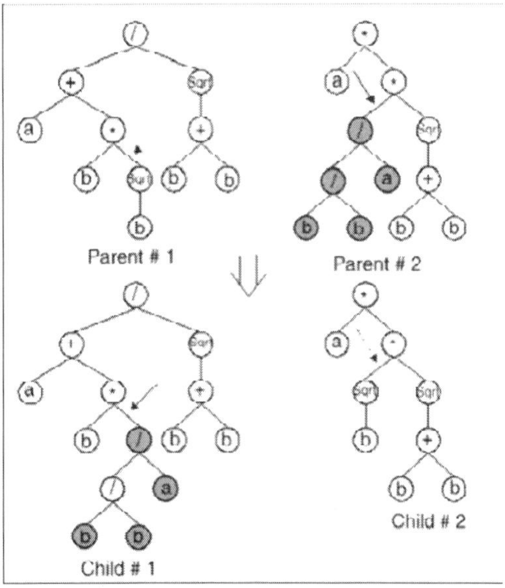

Fig. The Effect of a Crossover Operation.

- Reproduction : Reproduction, these condmain operation, is performed by simply copying a selected member from the current generation to the next generation. An interesting note about convergence of the population using the two main operators is made by Koza ... in genetic programming, when an individual incestuously mates with itself (or copies of itself), the two resulting offspring will, in general, be different.... As before, the Darwinian reproduction operation creates a tendency towards convergence; however, in genetic programming, the crossover operation exerts a coun terbalancing pressure awa y from convergence. Th us , convergence of the population is unlikely in genetic programming.

- Mutation: With genetic algorithms, mutation becomes an important operator which provides diversity to the population. As Koza points out, however, mutation is relatively unimportant in the new environm ent, bec ause the dyna mic siz es and shapes of the individuals in the population already provide diversity, and as stated above, the population should not converge. Thus, mutation can be considered as a variation on the crossover operation.

ASSIGNING A FITNESS VALUE

Now that we are basically left with two operators to implement, how do we select the individuals for crossover and reproduction? We have said nothing so far as to indicate how this done. Fortunately, genetic algorithms show us the way: fitness.

We must determine how good the individuals are at solving the given problem. And as with genetic algorithms, the crossover and reproduction operations are separate from the actual evaluation of the fitness, making the genetic programming operators problem-independent.

The measurement of fitness is a rather nebulous subject. Since, it is highly problem-dependent, we consider massaging the results to make fitness evaluation much easier, through a process known as scaling. Simply put, scaling standardizes the measurement of how fit a particular individual is with repsect to the rest of the population.

Based on the fitness value, we go about this selection for survival in one of two ways:

- To choose the individuals with the highest fitness for reproduction. "Only the strong survive."
- To assign a probability that a particular individual will be selected for either reproduction or crossover. This the method of choice, because it allows for more diversity. Some weak individuals may contain branches of code which are strong.

The fitness function is determined subjectively. For example, we could include the depth of the tree as a potential quality we wish to control, and therefore we could develop a fitness function which takes this into account.

❒

4

Tissue Culture Strategies for Genetic Improvement of Bamboo

Bamboos are among the economically most important plants world-wide. In Europe bamboos are used as ornamentals for gardens, but there is increasing interest for uses in ecological applications and as energy crops. Biotechnological techniques, including tissue culture, in vitro hybridisation, molecular markers and genetic transformation are crucial for the future of bamboo.

Micropropagation of bamboos has allowed developing a new type of ornamental bamboo that can be produced yearround with a high quality/ price ratio and distributed far more widely than classically propagated ornamental bamboos. Molecula r markers are us ed in quality control procedures.

Flowering of bamboos is still one of the greatest mysteries in botany, and breeding systems are non-existent. However, flowering can be induced reproducibly in tissue culture, both in seedlings and in adult bamboos, providing the only method for hybridisation. The flowering structures that are used are pseudospikelets, morphological features unique to the subfamily of Bambusoideae. These special propagules can be used for propagation, long term storage, for hybridisation and for genetic transformation. While flowering can be induced, controlled and reversed in tissue culture, a more fundamental approach to unravel the mechanisms of flowering include studies of cell division patterns and profiles of volatile components.

Some applications of biotechnology for bamboo are presented, with emphasis on research strategies in a SME. Furthermore, all the techniques developed are of use not only in horticulture, but also in agriculture and forestry worldwide.

Bamboo is an ornamental plant, mainly for garden use in Europe.

Besides, two different pathways for valorisation of biotechnological research have emerged:

- In th e tropics bamboo is a ve ry important plant, providing livelihood for over 500 million people and providing housing and shelter for over 1 billion people;
- "Bamboo for Europe", an EU- funded research project has allowed to develop the potential of bamboo as an agricultural plant in Europ with possible applications in wood industry and as energy crop.

So, any biotechnological developments for the improvement of ornamentals will also be useful on a much larger scale. To develop a commercially feasible process of bamboo propagation and genetic improvement three major components are important:

1. Fundamental research on bamboo physiology and genetics, including the development of tissue culture techniques for hybridisation of bamboo;
2. The development of axillary branching into a universal technique with a high clonal fidelity and very high efficiency;
3. Forward integration to optimise the added value of micropropagation.

MICROPROPAGATION OF BAMBOO

RESEARCH ON BAMBOO MICROPROPAGATION

For bamboo different propagation techniques are available, such as seed propagation, clump division, rhizome and culm cuttings. But these methods suffer from serious drawbacks for large or mass scale propagation. For mass scale propagation (> 500 000 plants per year) classical techniques are largely insufficient and inefficient, and tissue culture is the only viable method. Indeed, the order of magnitude of the demand for bamboo planting materials indicates that micropropagation will inevitably be necessary for mass scale propagation.

By now a large number of papers on micropropagation (the use of tissue culture for propagation only) of bamboos have been published, original papers as well as reviews and some in which tissue culture is described in a more general aspect. Many researches have focussed on somatic embryogenesis of seedlings of tropical bamboos. An inventory shows that at least 21 labs in South-east Asia were involved in bamboo tissue culture, mainly in India. Major research focused also on the clonal propagation of elite genotypes, either juvenile or adult. The number of papers about this subject however is much less, and this is solely due to lack of success. Indeed, technically the propagation of adult plants via axillary branching is much more difficult than with seedlings of tropical bamboos.

For tissue culture of bamboo the use of starting material (seeds or adult plants) and the choice of the propagation method are crucial. The two major advantages of using seedlings are that seedlings establish a new generation, and that the techno logy is easier.

But the disadvantages are considerable:

- Insufficient or no knowledge of genetic background,
- Restricted availability of seeds for most species and rapid loss of germination capacity,
- Comparison of in vitro to in vivo performance has not been thoroughly evaluated.

In addition there is a huge variability in responsiveness in tissue culture.

When using adult bamboos main problems are:

- Endogenous contamination,
- Hyperhydricity and instability of multiplication rates,
- Many problems with rooting also in bamboos that root readily in nature.

Rooting percentages for adult bamboos ranged from very low percentages of 10 per cent for Bambusa vulgaris to 73 per cent for adult Dendrocalamus longispathus. A rooting percentage of 77 per cent was obtained for adult Dendrocalamus giganteus in 3 or 4 weeks. But, while 77 per cent of success is good on 500 plants in a laboratory experiment it still represents a loss of 33000 when we transplant 100 000 plants. Low rooting frequencies are the major bottleneck to developing commercially viable protocols. The combination of photomixotrophic in vitro multiplication and photoautotrophic in vitro rooting stages resulted in improved transplanting success. Improvement of rooting percentages and transplanting has been achieved in various commercial laboratories.

COMMERCIAL MICROPROPAGATION OF BAMBOO

Micropropagation via tissue culture thus attracted a lot of attention but the translation and transformation of these expectations into commercially viable propagation systems has been beset with a number of problems that were either technological, or related to marketing. At present only a very limited number of laboratories specialise in the production of bamboos at a commercial scale. Oprins Plant, Belgium, currently has over 60 genotypes in culture initiated from adult genotypes of which about 30 on a commercial scale.

Species and varieties of Arundinaria, Chimonobambusa, Fargesia, Phyllostachys, Pleioblastus, Sasa, Sasaella, Semiarundinaria, Shibataea and Yushania (temperate bamboos) and Bambusa, Dendrocalamus, Dinochloa, and Thyrsostachys (tropicals) a reproduced commercially. Rooting percentages of most species are between 85 and 100 per cent, depending on species and season of transplanting (unpublished results). These bamboos are produced either as ornamental, or for tropical forestry. Other laboratories focusing on bamboo tissue culture include West Wind Technology, USA, and Bamboo World, Australia.

Several other laboratories such as Piccoplant, Germany, and Microflor, Belgium, produce a more limited number of bamboo species as part of their tissue culture operation. Other laboratories initially involved in tissue culture of bamboos however, no longer produce bamboo commercially, for example TERI,

India and Thai Orchids Lab, Thailand. The main problem is that micropropagation of bamboos also has to deal with the development of new markets and market opportunities. Logistics and supply chains such as in ornamental horticulture where tissue culture labs are separate entities seem not to work. Instead the laboratories have to develop their own markets, marketing and sales for bamboo plants. This strategy of forward integration needs to go beyond the classical scheme of tissue culture.

The tissue culture phase terminates at Stage III, but in regard to marketing of plants one can also distinguish subsequent stages:

- Stage IV, the transplantation stage with the end product being rooted plantlets in trays,
- Stage V, the production of liners, either for production of saleable plants or for use as micromotherplant,
- Stage VI, the production of saleable plants.

This distinction is important if the complete chain of production is integrated in a single company, since, this determines the added values. In our case the laboratory produces ornamental plants exclusively for Oprins Plant's own nurseries. A balanced production and logistic chain, where bamboos are produced in Belgium and grown in the nurseries in Spain and France, has allowed to produce high quality ornamental bamboos year round at prices which are 40-60 per cent below the current market prices for bamboo.

These plants are currently being sold under the brand name BambooSelectTM, which is an additional tool in marketing. In an agricultural perspective it is now possible to produce the same genotypes for large scale plantations in Europe at prices comparable to other energy crops.

MONITORING GENETIC STABILITY IN BAMBOO TISSUE CULTURE

AFLP Markers in Quality Control of Micropropagation

Somatic mutations in bamboo are highly valued in horticulture. In various bamboos somatic mutations are known that alter the stem colour or stem shape, such as Bambusa ventricosa with inflated internodes. But not all are stable in tissue culture.

The bulbous internodes of Bambusa ventricosa were lost after tissue culture and variegation of stems and leaves disappears very often in tissue culture. Tissue culture has been associated with molecular aberrations in general (including also chromosomal defects), and phenotypic defects may show up only some years later. Therefore it is important to devise methods of quality control at the tissue culture stages to minimise the potential danger of such defects, for exa mple when plants of Stage VI are sold in large quantities.

Generally it is very difficult to find precise defects, especially when somatic mutations are possibly linked to transposon activity. In a first approach AFLPTM markers are used to assess the stability and clonal fidelity of tissue cultured bamboos.

In one experiment Phyllostachys species and cultivars were collected:

- From the garden of Dr. Jacques Van Dooren, Kumtich, reference collection of bamboo for Belgium
- From the tissue culture lab at Oprins Plant
- From plants transplanted in the greenhouse.

Using this set up, it should be possible to detect if genotypes in the collection and in propagation have the correct names, and whether the plants cultured are subject to considerable clonal variation.

The most bamboos in tissue culture and transplanted in the greenhouse group together with the collection plants. The exception is Phyllostachys vivax from the laboratory that turns out to belong to the nigra group, while the collection plant of P. vivax does not group at all with the nigra's.

This is a clear case of mistaken identity, due to mislabelling of the motherplants obtained from China. If one observes the dendrograms it is seen tha t all P. aurea cultivars, all P. nigra cultivars, and all P. aureosulcata cultivars group together. They are, according to this test, 100 per cent identical. However, they differ in stem colour, leaf colour, habit etc, and these differences can be observed readily in the field. This also means that the method used is not sensitive enough to detect those differences. But it clearly shows that AFLP is a useful method when used as part of quality control in the tissue culture process, to assess correct clonal identity. This can then be used towards customers. AFLP markers are now used routinely besides other measures to ensure clonal fidelity. These other measures include the use of axillary branching as propagation method and annual initiation of new cultures, so that the maximum number of subcultures is 12 to 14.

MOLECULAR BASIS OF SOMATIC MUTATIONS IN BAMBOO

Based on phenotype and mode of action somatic mutations involved in stem colour of bamboo are possibly caused by transposon activity. The study of transposon activity and their possible relation to certain mutations is very difficult. In bamboo no breeding systems are available to detect transposon activity through classical breeding. Molecular approaches may provide some shortcuts, but to link the presence and activity of transposons to certain phenotypic markers is again a difficult task.

A first approach to identify the molecular basis of these somatic mutations was to use primers based on sequence information from the 4.5 kb Ac9 transposon from maize. Several primer combinations were used and several copies with considerable homology to the original Ac9 transposons were present, which eventually turned out to be a useful and relatively cheap method to identify bamboos, at the species level.

Primers designed to detect Activator-like transposons in Petunia could also be used to detect fragments in bamboo. The same fragment was detected in the three bamboos tested, namely Bambusa vulgaris, Sasa veitchii and Phyllostachys edulis. The fragment of Bambusa vulgaris (hATbv1) was sequenced and shown to be homologous to members of the hAT superfamily of transposons, to which also Ac (corn), Tam (Antirrhinum majus) and hobo (Drosophila) belong. The homology was as high as 60 per cent in some regions, which conforms also to earlier findings of Ac-like sequences in Bambusa multiplex. It is very likely that many other transposons are present in the bamboo genome as well.

A next step is to use these AFLP-primers in combination with methylation sensitive AFLP, in which methylation sensitive restriction enzymes are used to cleave s ites within the ba mboo tra nsposon . Th e rela tion betw een methylation patterns and transposon activities is well established and the activation of transposons during tissue culture is also a well documented phenomenon.

The use of genotypes that have been cultured for six, eighteen, twenty- four and thirty-six months should allow to assess if long term tissue culture has any influence on methylation in general and on methylation patterns of transposons. This research is currently ongoing.

FLOWERING OF BAMBOO IN TISSUE CULTURE

Because of the peculiar flo wering habits in bamboo it has been almost impossible to breed for superior traits in woody bamboos. The first reports on tissue culture flowering of bamboos caused great excitement. It opened up the possibility of controlled flowering that can be used for breeding of bamboo. Since, then in vitro flowering has been observed in many types of bamboos, both in seedlings and mature bamboos. More than 10 years after the first report on in vitro flowering in bamboo some positive results have been obtained, but practical and commercially exploitable results have not been reported yet. The most promising potential is for hybridisation under methods really become applicable at agricultural scales.

Recently a new research project was initiated in our group to improve the induction of flowering and the quality of flowering parts, in order to be able to hybridise bamboos in the future. While this project aims mainly at developing techniques, it also includes various approaches on a more fundamental level. The use of pseudospikelets is central in this research.

Pseudospikelets occur only in bamboos, not in other grasses. In grasses a spikelet bears two empty glumes, but the glumes of pseudospikelets in bamboos subtend dormant buds that can develop into new pseudospikelets. The proximal parts of the pseudospikelets do not develop into flowers but in tissue culture conditions keep on multiplying indefinitely, allowing to establish monocultures of pseudospikelets. Flowers are formed in the distal parts of individual pseudospikelets and do not develop to anthesis under normal tissue culture conditions.

They can however, be forced to develop fully under more adverse conditions. Cultural conditions can control flowering in vitro to a large extent. Factors used for the induction of pseudospikelets are mainly cytokinins. However, cytokinins are involved in various flowering processes and more fundamental research is urgently needed. Unfortunately, despite the huge potential of tissue culture flowering of bamboo, to our knowledge only one other laboratory is actively engaged in this field of research.

Despite the presence of flowers in the axils of pseudospikelets, dormant buds can develop into vegetative shoots. Pseudospikelets can also be regarded as highly contracted axes, in which leaves are reduced to glumes. Instead of propagules and plantlets of 2 to 5 cm in tissue culture, pseudospikelets have lengths between 0.25 and 1.5 cm. This is a miniaturisation of the plants and allows a good opportunity to use these pseudospikelets for propagation of bamboos.

With these smaller clumps, operators can cut and transfer much easier, while the mode of propagation is still via axillary branching, thus the method with lesser risk for somaclonal variation. Since, leaf area is much reduced, quality problems in the propagation stage are much less important. This allows for a considerable cost reduction in micropropagation, since, operator costs represent 75- 80 per cent of cost price of tissue culture plants.

In tissue culture systems for propagation flowering of bamboo represents considerable risks since, many bamboos flower monocarpically in nature and die completely after flowering. It is very important to be able to control flowering and reversion of flowering of bamboo. Therefore the influence of tissue culture flowering on vegetative growth post vitro has been studied extensively in three independent experiments (unpublished results).

- Plants of Bambusa tuldoides regenerated from pseudospikelets grow as vegetative plants after hardening;
- In large scale vegetative propagation only two flowering plants of Phyllostachys have been observed after transplanting about 500000 plants,
- Tests with Sasa palmata have shown that even when flowering plants are transplanted in the greenhouse, these revert to the vegetative state within a few weeks.

These tests have shown that tissue culture technology does not lead to monocarpic flowering of saleable plants. Using pseudospikelets as propagules we have now a system which can be used for propagation and for the induction of flowering and flowering structures. The same propagules can also be used for long term storage and for genetic transformation.

❐

5

Genetic Resources and Plant Breeding: Uses of Marker aided Selection

In order for genetic resources to be efficiently utilized in plant breeding programmes, it is first necessary to determine whether useful genetic variation exists in the material and secondly to develop the most cost-effective method of introducing the potentially useful genes into commercially acceptable material. Until recently, genetic analysis has relied on conventional segregation analysis in controlled crosses for qualitative traits determined by major genes, and on standard biometrical methods for quantitative traits controlled by polygenes. Now, however, the use of DNA techniques has opened up new possibilities for genetic analysis and breeding, and in this chapter the general principles and applications of these molecular marker-based procedures will be discussed.

In its specific sense, marker-assisted or marker-aided selection (MAS) is the use of appropriate easily recognizable genetic markers to facilitate or accelerate the selection of linked genes controlling useful agronomic traits. The nature of the marker loci themselves is not important and they are used simply to indicate the presence of useful alleles of genes of commercial or practical importance. However, the term has also come to include the wider use of markers to improve breeding programmes.

In an ideal world, selection should be aimed directly at those genes which control the trait to be improved, but this assumes that genotypes at these target genes can be recognized easily, unambiguously and at an appropriate time without impeding the breeding programme. There are several reasons for using marker genes to improve selection efficiency.

Difficulties in Identifying Trait Genotype

The genotypes at a single-trait locus may be difficult to recognize because the desired alleles may have poor penetrance, be recessive, or interact with other genes or the environment. Environmental variation itself may impede accurate genotyping by causing the phenotype of different genotypes to overlap; this is particularly a problem with polygenic traits.

Some phenotypes such as resistance to a particular pest, pathogen or abiotic factor, like drought or salinity, may only appear under particular conditions which are difficult to define or control. These difficulties may be more severe during the early stages of a breeding programme when plant numbers are too few to allow adequate replication or the breeder does not want to place his valuable yet scarce breeding material at hazard.

Earlier Selection

Another major reason why it may be valuable to use marker or indicator genes is to reduce the time from sowing to selection. Many traits of economic importance are only apparent in the mature plant and so may not occur for months or even years after sowing. This is particularly a problem with biennial species and long-lived crops such as trees. Indicator genes, on the other hand, may be detectable in seedlings within days of sowing, if not in the seeds themselves, so avoiding the waste of valuable resources involved in raising and harvesting plants, most of which may prove to be of no value to the breeding programme.

More Intense Selection

Selection at a juvenile stage, particularly among seedlings, may also allow much larger populations to be studied and hence, more intense selection to be applied. Indeed, selection may be possible in tissue culture without raising plants or running trials.

Nondestructive Scoring

Many characters are scored before maturity and the act of scoring may prevent seed being collected. Imposition of selection for pests and diseases may result in plants which are less able to produce seed. The use of certain types of marker loci may only require the removal of small quantities of leaf or other material to permit full genotyping of the plant.

Linkage Drag

Conventionally plant breeders have used backcrossing with selection to introduce a useful allele, such as one conferring disease resistance, from one strain into another. The source of the allele is often an alien or exotic species, but may be a related cultivar. The F1 is backcrossed to the recurrent parent, and backcrossing is repeated for five to ten generations, the intention being to introduce the desired donor allele whilst returning to the recurrent parent's genotype at all other loci. At every backcross generation, the breeders elects for the phenotype of the allele to be introgressed whilst simultaneously selecting for the phenotype of the recurrent parent for all other traits.

It was shown by Stam and Zeven (1981) that this procedure results in the leaving of very large amounts of donor genome associated with the selected alien allele. On average, about 20per cent of an average chromosome will consist of a donor segment even after ten backcross generations, while in any given line the actual size could be much more. Thus, as well as introducing useful alleles, the breeder will also be introducing very many undesirable alleles from the donor and so the gain in one trait may be at the expense of losses elsewhere. The use of marker loci can enable the breeder to reduce the size of this unwanted associated region as well as to accelerate the speed of return to the desired genotype of the recurrent parent.

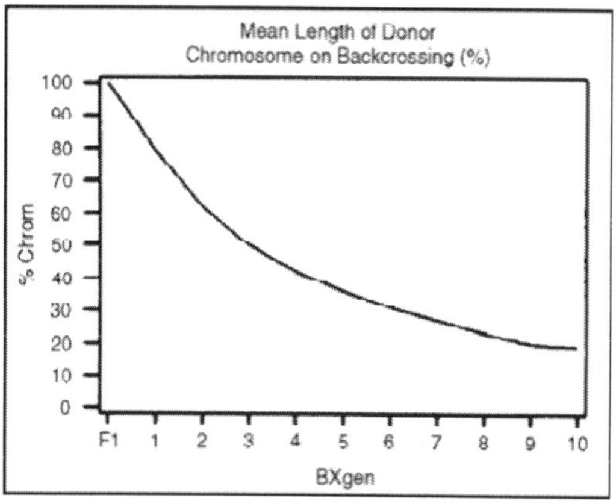

Fig. Linkage Drag.the Average Amount of Chromosome Surrounding a Selected GeneFollowing Backcrossing.Bxgen=Generation of Backcrossing from F1

Heterosis Potential

Heterosis, or hybrid vigour, is due mainly to the presence of dispersed dominant alleles in the parental cultivars. Breeders concerned with developing lines for hybrid production are often concerned to identify lines which differ at as many gene loci as possible. It is often assumed that this implies that the lines also have very different genotypes at marker loci and hence, such loci can be used to assess the genetic distance between potential parents, but this can be a dangerous assumption unless a wide range of different types of markers are used.

Quality Control

A major problem faced by most conservationists and breeders is to maintain the integrity of the material they are working with and to keep it free from contamination. Contamination can occur in various ways. It may arise through errors due to carelessness in labelling or sorting seed or plant material, mistakes in maintaining the pedigree, or as a result of outcrossing due to stray pollen from unrelated material. Such errors are difficult to recognize when only the gross phenotype is available as a guide, but genetic markers can provide a clear genetic fingerprint which can be used at any time to confirm the origin of the material. It can also be used to confirm that the intended cross or self has occurred, or to separate the desired progeny from a particular parent which might be a mixture of selfs and crosses.

Other Uses

Marker information can be used in other situations also: distinguishing haploids, diploids and aneuploids, for example, particularly following wide interspecific crosses or tissue culture. Doubled haploid lines produced by microspore or ovule culture can be recognized in this way and possible aneuploids arising during chromosome doubling can be identified and removed. Not only can genetic markers identify contaminant material but they also allow closely related or identical material to be identified in genebank collections, so reducing unnecessary duplication.

TYPES OF GENETIC MARKERS

Up to the mid-1960s the only easily usable genetic markers were those that produced clearly visible effects on the plant's phenotype such as colour, size, shape, disease resistance, etc.. All these were the result of mutations to alleles which had, obviously, a major effect on the development of the plant. Such mutants were rare and generally not desirable in commercial material although they may be selected by breeders of decorative plants for their novelty value. As genetic markers, however, they have many disadvantages. They often affect the fitness of the individual and could well have effects on important agronomic traits, i.e., they exhibit pleiotropy. Because they are rare, it would be necessary to bring several such mutants together in the same material in order to use them, and they would not normally be present in commercial cultivars or, indeed, in wild material. This not only takes time, but the combined effects of many such alleles would mitigate against their value as a genetic tool.

In the 1960s through to the 1980s, naturally occurring genetic polymorphisms at the protein level came to be easily recognized. Variation in enzymes and storage proteins was detectable on an electrophoretic gel for a very high proportion of gene loci which could be studied in this way and, unlike major mutants, such naturally occurring allelic variants had much smaller effects on fitness and hence, were more common and existed naturally in populations. The number of such polymorphisms was, however, insufficient to pr ovide th e covera ge of the ge nom e de sired by br ee de rs an d conservationists. The development of techniques during the late 1980s up to the present revealed the vast amounts of genetic polymorphism which exists at the DNA level and has revolutionized genetic analysis, opening up a whole new range of genetic tools.

As described elsewhere in this book, this variation arises through the existence of occasional base changes in the DNA which can be recognized by restriction enzymes or by primers used in analysis involving the polymerase chain reaction (PCR). Much of the DNA, possibly as much as 90per cent in many species, is noncoding and henc e, relia nce on morph ological or biochemical variants was only using the restricted variation that plants could tolerate within the coding parts of their DNA.

Clearly, there are limits to the tolerance of such variation. Variation in the noncoding regions, whether in the intergenic regions or within introns, is probably under far less constraint by natural selection, and hence, the number of polymorphic sites is potentially enormous. Because most of these sites are outside coding genes they are referred to as loci rather than genes; genes are coding regions whilst most marker loci are not.

Restriction fragment length polymorphisms (RFLPs) have been the most widely used until now. They have the advantage of being codominant and hence, all genotypes in a cross can be identified. On the other hand they require quite large quantities of DNA, and hence, plant material, and generally rely on radioactive probes, although fluorescent techniques are becoming more popular.

As a result, RFLPs are expensive to use. The use of radioactive probes incurs safety considerations, as well as taking longer to visualize the bands. PCRbased markers are now being used more widely.

Randomly amplified polymorphic DNAs (RAPDs), for example, require far less material because only small amounts of DNA are needed and the required sequences are then amplified. They do not involve radioactive probes and are both quick and cheap. On the other hand RAPDs are dominant although it is possible to distinguish heterozygotes from homozygotes by the amount of product. Amplified fragment length polymorphisms (AFLPs) offer even cheaper and more easily identified and extensive PCRbased polymorphism.

Because of the number of polymorphisms which exist on individual gels and the speed of production, computer software is necessary to scan the banding patterns on the gels and to analyse and assimilate the information that is generated. Other popular markers are microsatellites (short sequence repeats) and cleaved amplified polymorphisms (CAP)-based primers.

All the genetic variants discussed above will be subsumed under the broad title of marker loci. They may be of little interest in their own right, and this is particularly true of the molecular polymorphisms; their value lies in their use as markers of more useful genes.

GENE MAPPING

An important, though by no means an essential, step in genetic analysis is to produce genetic maps of the marker loci. Such maps are often referred to as 'framework maps' because they provide a framework within which important genes can be located, as well as providing a means of comparing chromosome organization in other closely or distantly related species.

There are two stages to mapping. Firstly, to arrange the markers in a linear sequence separated by an appropriate map distance, i.e., to construct a linkage map. The second is to relate the linkage maps to particular recognizable chromosomes. The latter is often the most difficult and generally not essential for breeding or conservation work. We will, therefore, concentrate on the former.

Chromosomes contain a single linear molecule of DNA and hence, the markers on that chromosome occur a t particular positions along that molecule. Typically, each chromosome contains 107 to 108 base pairs (bp), i.e., 104 to 105 kilo-base pairs (kbp) of DNA, while a typical structural gene, coding for a polypeptide chain, would be between 1 and 2 kbp long. Assuming only 10per cent of the genome is coding DNA, then a chromosome probably contains something of the order of 1000 to 10 000 genes. Fortunately, the fact that the chromosome is a linear molecule means that the genes need to be mapped in only one dimension.

Currently, the only useful method of gene mapping, at least as far as breeders and conservationists are concerned, relies on recombination between homologous chromosomes resulting from genetic exchange, which can be seen as chiasmata at diplotene through first metaphase of meiosis. A single chiasma on a particular chromosome results in half the gametes from such a meiosis being recombinant for that chromosome: half the gametes will contain a copy of that chromosome that is part maternal and part paternal in origin, while the remaining gametes will be entirely parental. Very few plant species have a sufficiently low number of clearly recognizable chromosomes that the number of chiasmata on a particular chromosome can be compared in different nuclei.

However, in many species the total number of chiasmata in each diplotene nucleus can be counted, at least in pollen meiosis, and the average number of chiasmata per nucleus calculated. The number and position of chiasmata on a particular pair of homologous chromosomes will vary from nucleus to nucleus; it is as though there are a large number of potential sites of exchange, but that in any given meiosis only a few sites are actually involved.

The longer the chromosome the more potential sites there might be. There will invariably be one chiasma and there may be two, three, four or more; a typical set of results is shown in Table. A chromosome that has one chiasma on average is said to be 50 centimorgans (cM) long – a map length unit named after the American geneticist Thomas Hunt Morgan.

50per cent recombinant chromosomes. By extension, a chromosome with an average of 2.5 chiasmata is said to be 125 cM (i.e., 2.5 350 cM) long. It appears that, as a rough rule of thumb, each chromosome has on average two chiasmata and so is 100 cM long. It follows, therefore, that providing the haploid chromosome number is known (n), the total map length will be approximately 2 3 n 3 50 cM. The average number of chiasmata per nucleus has been calculated in many species and this rule generally holds.

Table. Chiasma Fraquency and Mapping.Chiasma Frequency in Chromosomes of Secale Cereale.

Number of chiasmata per chromosome	Frequency	Percentage unrecombined Chromosomes
1	0.267	50.0
2	0.716	25.0
3	0.017	12.5
Mean	–	22.0

This rough rule is important because it gives the geneticist an idea of the total genetic map length that should be expected as more genes or markers are mapped. It provides a guide to the extent to which the currently mapped markers cover the full genetic map. Because the map is based in chiasma units, it is not the same as a physical map. The distribution of chiasmata is not uniform over a given chromosome in a species and will vary with the chromosome and the species.

Although a knowledge of the number of chromosomes and, ideally, the mean chiasma frequency per nucleus for the species, indicates the total map length, the actual map has to be constructed from examining the frequencies of progeny in crosses; i.e., maps are constructed from information based on the consequences of chiasmata, namely recombination of genetic markers. Recombination of genetic markers is identified by the presence of gametes that contain recombined markers, and such gametes can be recognized from the phenotypes of progeny of a cross.

The frequency of such gametes is an estimate of the recombination frequency between the two markers concerned. The closer two markers are to each other on the chromosome, the less likely a chiasma will occur between them and the less likely they are to recombine. Providing the two markers are sufficiently close that either none or one chiasma occurs between them but never more, then the recombination frequency as a percentage is equal to the map distance in centimorgans as described above. The problem arises when the markers are sufficiently far apart for two or more chiasmata to occur between them in some meioses. It can be shown that not only does a single chiasma between a pair of markers result in 50per cent recombination between them, but so also do two, three or more chiasmata, on average.

It is necessary to say 'on average', because there are a variety of possible consequences with two, three or more chiasmata in any given meiosis but, in practice, when one is looking at the progeny of a cross or self, each progeny will be the result of gametes from different meioses.This fact implies that although map distance increases linearly with the number of chiasmata, the same is not true with the frequency of recombination, which can never be more than 50per cent for a pair of markers even though they might be 100 cM apart at opposite ends of a chromosome.

These relationships are illustrated in Fig.. In order to overcome this problem, mapping functions have been devised to correct for multiple chiasmata. The most common of these are the Haldane (1919) and Kosambi (1943) mapping functions. Haldane's assumes that the probability of none, one, two or more chiasmata in a given interval follows a Poisson distribution, i.e., that the chiasmata are independent, random events. Kosambi's method allows for a certain degree of nonindependence in chiasma occurrences. It has long been known by cytogeneticists that chiasma interference occurs over quite large regions of a chromosome.

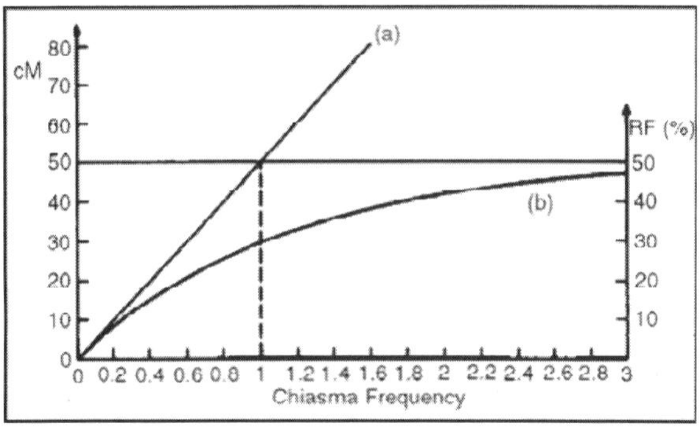

Fig. The Relationship of Chiasma Frequency with (a) Genetic Map Length and(b)Recombination Frequency.

Interference means that if a chiasma occurs at a particular point on the chromosome, the next one will never occur closer than some fixed distance, the interference distance, from it. Beyond this distance, there is a short distance in which the interference disappears and the next chiasma can then occur randomly outside this range.

Observational data in various species of plants are confirming this and suggest that the interference distance is generally between 15 and 20 cM, i.e., 15 to 20per cent of a typical chromosome on either side of the chiasma. This has important and useful consequences for genetic and breeding work as will be shown later.

Recombination frequencies between pairs of markers can be scored in a wide variety of different crosses. The simplest to use are generations derived from an F1 because only two alleles are segregating, and their distribution generations that can be used are F2, backcrosses (Bc), recombinant inbred lines (RILs) or doubled haploid (DH) lines, with F2 s be ing the most informative.

General formulae for calculating recombination frequencies are given by Allard (1956). In outbreeding species where F1s are not available, a given individual can be considered as an F1 and its selfed progeny as being an F2, even though the inbred parents do not exist. If it cannot be selfed but controlled crossing to another individual is possible, then again those genes segregating in the cross can be mapped although the situation is more complex.

There could be from two to four alleles segregating at each locus and, with respect to any one locus, the cross could represent an F2, Bc1 or Bc2. Moreover, the distribution of alleles in the two parents cannot be known and has to be inferred from the progeny. This complexity is compounded by the large number of marker loci that may be segregating in any given cross.

Fortunately, a range of software packages are available to estimate these recombination frequencies, identify linkage groups, assign the markers to the most likely order and space them in map units (cM) on these linkage groups; examples are MAPMAKER and JOINMAP. Ideally the number of linkage groups should be equal to the chromosome number in the gametes but, unless the markers available provide a good coverage of the genome, it is likely that those markers on a given chromosome may appear as two or more separate linkage groups simply because the subsets are not sufficiently close for them to be recognized as being together on one chromosome.

A typical marker framework map is shown in Fig. for the chromosomes of Brassica oleracea. Clearly these maps provide a very detailed coverage of the chromosome. Many of the markers are very close, i.e., less than 10 cM, and over this distance it matters little which of the two mapping functions is used. With the more widely spaced markers, i.e., recombination frequency >15per cent, Haldane's function will exaggerate their distance apart because it will allow for more double crossovers than actually occur. As more markers are mapped, the total length should converge on the value predicted by the chiasma frequency, i.e., approximately 2 3 n 3 50 cM. It is important to remember that genetic map distances have standard errors which are dependent on the size and type of population used to construct the map, the mapping population. The map obtained is that which best fits, given the data and the assumptions underlying it, but there may well be other maps which also fit the data well, though are slightly less likely. Any particular map always develops a greater aura of respectability once it is published, frequently without stating its reliability, and subsequent yet different maps are treated with suspicion. A recombination frequency of true value p has a standard error of ?[p(1 2 p)/N], where N is the size of the gamete population. For example, if p is 0.1 (i.e., RF = 10 per cent) and N = 100, the standard error is 0.03 or 3 per cent. In other words the 95 per cent confidence interval of the estimate lies approximately between 4 per cent and 16 per cent.

There are other reasons, apart from statistical sampling, why maps could be wrong. The most obvious is the accuracy with which the data are scored and recorded.

Autoradiographs and banding patterns on gels can easily be misread and research workers are loath to discard data even though its interpretation is ambiguous. All data should be checked independently by two or more people and any ambiguities removed. Wrong scores will bias the data and often exaggerate map lengths as they suggest double recombination; wherever double recombination appears to have occurred in two short, adjacent intervals, the data should be checked. Changes in methylation rather than scoring errors could cause a restriction site to appear or disappear so giving the false impression of double crossing- over, or it could simply be an error.

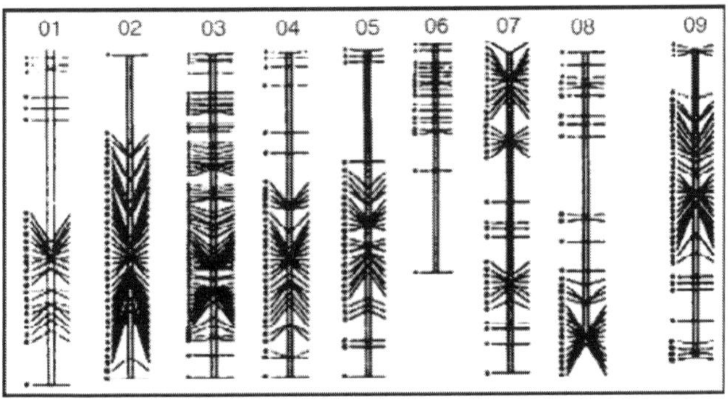

Fig. Typical Molecular Marker Framework Map. The Nine Groups of Brassica Oleracea Based on RFLP, Isozyme and Morphological Markers

Where maps appear to have major inconsistencies in different crosses, different chromosomal structural arrangements may be responsible, such as translocations or inversions. There is considerable evidence that chiasma frequencies, and hence, recombination frequencies, may be quite different in male and female meioses even on the same plant although in other studies this may not be so.

Thus although the order of markers on a linkage group may remain the same, the relative distances between them may be quite different if the map d erives from m ale versus fem al e mei ose s. For ex a mple , in Brassica it appears that the genetic map obtained from a backcross where the F1 is the female parent is 60per cent longer than when the F1 is the male parent. It is also clear that environmental factors during meiosis, particularly temperature, can a ffect the number and distri bution of chiasmata, and this could be very critical if the crosses are set up at different times of the year.

Two or more different populations will almost certainly be segregating for different combinations of polymorphic markers, although some at least should be in common. These common ones can be used to overlay the two or more maps and a consensus map produced from an amalgamation of these. Again software is available to produce these consensus maps. Despite the various error-causing factors discussed above a considerable degree of consensus can be produced from such populations and, considerable similarity of chromosome sequence is conserved between even distantly related species, allowing the potential for cross-species genetic transfer.

It was stated above that the total map length should be dictated by the chiasma frequency, in so far as this can be accurately measured. When the first few genes were initially put onto genetic maps in the first 40 years of this century, they only covered a small part of the genome – except in wellstudied species such as Drosophila and maize. As more and more genes were placed on the map, so the total map lengths increased towards the asymptote dictated by the chiasma frequency. However, during the first decade following the use of molecular markers, the map lengths of some species appeared to exceed that expected, and it was even argued by some that the chiasma theory of recombination might be wrong.

No one would wish to argue that chiasma frequency data are free from error but there would have had to have been quite excessive underscoring to result in the disparity which was occasionally found. It would appear, however, that the excessive lengths were due in part to errors in scoring and to the use of small mapping populations, because subsequent map sizes have shown a progressive decrease towards the predicted asymptote. Scoring errors bias estimates of RF generally upwards, while RFs have to be large to be detectable in small populations. Our own experience with mapping in cereals and brassicas has also shown a progressive decrease in estimated map length as more markers can be found to fill some of the large gaps in the map and as errors are removed from the mapping data.

LOCATING GENES OF MAJOR EFFECT

The use of extensive molecular and other markers as described above provides a general framework map. Although this has value in its own right for comparing linkage groups in different species as well as possible structural variation within a species, the main value lies in providing a set of markers to locate genes of economic or special scientific interest.

Locating individual major genes is a straightforward development of the procedures above. Let us assume that a cultivar is identified which contains an allele of interest which changes the phenotype in a clearly recognizable fashion and which appears to segregate as a single gene in crosses to other cultivars. Using a knowledge of the positions of existing marker loci, a small subset of markers can be chosen which provide a reasonably even coverage of the genome, say every 20 to 30 cM, i.e., approximately five per chromosome. These then have to be shown to differ between the two cultivars and any which are monomorphic replaced by polymorphic markers.

A rough position of the gene of interest can then be obtained by bulked segregant analysis. This involves taking the mapping population, e.g., a Bc, and dividing the plants into two groups depending on whether or not they show the phenotype of interest. Bulk samples of DNA are taken from each group, and the two samples assayed for the marker systems chosen as in the paragraph above.

Any marker which is unlinked to the gene to be located will produce similar banding patterns in both samples because it will be segregating independently of the target gene. Conversely, should one of the markers be very closely linked to the target gene, it will co-segregate and the two samples will show quite different patterns for that marker.

In practice, complete co-segregation with any one marker is not very likely but, given a reasonable spread of markers over the genome, one or two are likely to show a strong association with the two groups. Thus although bands associated with both marker alleles will be found in each sample, their relative intensities will be quite different because of the association with the target gene.

Having located the chromosome and the region on the chromosome, the actual position can be identified in a segregating population using the principal marker associated in the bulked segregant analysis together with other polymorphic markers situated about 10 cM on either side of the principal marker.

The target gene is then mapped with respect to the three markers. This approach can be used for a wide variety of populations and traits. Even if the trait has poor penetrance, the affected group will clearly show the marker banding pattern even if the nonaffected group consists of a mixture due to misclassification.

MAPPING GENES CONTROLLING QUANTITATIVE TRAITS

Quantitative traits such as yield, quality, height and flowering time, which are controlled by several genes and are greatly influenced by the environment, create particular difficulties for gene mapping. The problems arise because the genotype for the trait concerned can never be clearly identified from the phenotype; many different genotypes could produce the same or very similar phenotypes whilst the same genotype can result in different phenotypes depending on what may be elusive factors in the environment.

Whereas with a single gene difference controlling a major effect, two or more Mendelian phenotypic classes might be recognized in ratios of 3:1, etc., quantitative traits resulting from the joint segregation of many genes show a continuous, often normally distributed range of phenotypes. The total phenotypic variation in an F2, for example, VP, is made up of genetic and environmental components, VG and VE. The genes underlying such traits are variously called polygenes, effective factors or, more recently, quantitative trait loci or QTL.

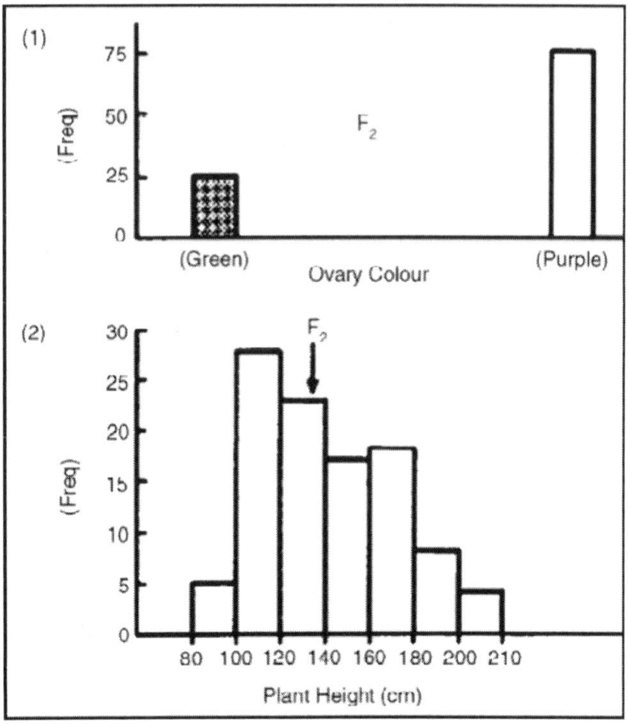

Fig. Contrast Between Distribution of a Qualitive and Quantitative trait in an F2. Data for NIcotiana Rustica,Ovary Colour and Height.

However, it is still possible to map and measure the effects of the QTL by techniques analogous to those used for single major genes, i.e., by looking for the effects of co-segregation of particular marker loci with differences in the quantitative trait. Various mapping populations can be used as before, F2, Bc, RIL, DH or open-pollinated full sib (FS) or half sib (HS) families, and the individuals or families are scored both for their phenotype for the quantitative trait(s) and for their genotype at the marker loci. The type of population that is used will depend on the traits studied and the breeding system of the plant species but, other things being equal, an F2 population will normally provide the most information for a given size. Unfortunately, most traits of economic importance such as yield, quality and disease resistance are not usefully scored in an F2 population because one is interested in how the material performs in something similar to the high-density monoculture that it will encounter in agricultural practice. Measuring yield in an F2 population with spaced plants is straightforward and statistically very powerful but it may well not provide any insight into how yield is controlled in agricultural practice.

It is therefore necessary to use a plot structure for most traits in an attempt to simulate commercial conditions. For this reason most plant breeders will prefer to work with genotypes that can be extensively and easily replicated such as RILs,

DHs or F3s. It is also easier to explain the principles underlying QTL mapping using a population of DH lines, and so for both reasons the methodology will be illustrated with DH lines, although the same principles apply to all populations. Because several trait loci are concerned, probably on different chromosomes, bulk segregant analysis is of limited applicability, at least initially.

THEORY OF QTL MAPPING IN DOUBLED HAPLOID LINES

Doubled haploid (DH) lines can be produced parthenogenetically from an F1 by microspore or ovule culture and, in some species, such as wheat and barley, by wide outcrossing. Each original DH plant is derived from a single gamete, which subsequently becomes homozygous and disomic either naturally or by treatment with the drug colchicine, which inhibits spindle formation, and hence, chromosome disjunction, at metaphase of mitosis.

Selfed seeds from such a plant produce a DH line of identical individuals so the line becomes effectively immortal. If we consider an F1 that is heterozygous for a marker locus M1/M2 and for a linked QTL locus Q1/Q2, where the 1/2 indicates the allele with an increasing or decreasing effect on the trait, this will produce the gametes and hence, the DH lines shown in Table.

If th e DH lin es are scor ed fo r some trai t f or wh ich th e Q+/ Q+ homozygotes increase the trait above the overall mean, m, by a units while the Q2/ Q2 homozygotes decrease the mean by a, then the means o f t h e f o u r g a me t i c t y p e s ar e as s h o w n i n T a b l e, wh er e R is th e recombination frequency between the QTL and the marker. From this it follows that the mean trait score of all those DH lines which are M1/M1 (M1M1) is m 1 a(1 2 2R), while the mean of all those DH lines which are M2/M2 (M2M2) is m 2 a(1 2 2R). In other words, half th e difference between these two means (M1M1 2 M2M2) is a(1 2 2R). Clearly, if the marker is so close to the QTL that it never recombines, R will be 0 and the marker difference will represent the QTL effect, a. Conversely if the two loci are unlinked, i.e., R = 0.5, then the difference will be zero. This effect is, of course, the basis of bulked segregant analysis. In general, the magnitude of the marker effect will decline linearly with the distance of the marker from the QTL in terms of recombination frequency, and so with several linked markers, their individual trait effects will be as shown in Fig.. Similar Arguments Apply to Other Populations Such as F2s and RILs.

Table. QTL model for market analysis in doubles haploid (DH) lines from and F1.(a)The F1 constitution; M=Market and Q = QLT genotype.(b)DH genotypes,frequencies and genetic values. (c) Expertations of marker means and differences.R=recombination frequency between M and Q; m and a are the mean and additive genetic effects of the QLT;d is the additive genetic deviation at the market locus

If the trait means of the markers and their map are known very accurately, and only one QTL existed on the chromosome, then as Fig. shows, it would be relatively easy to locate the QTL. Figure shows the distribution of flowering time among a number of doubled haploid lines of Brassica oleracea containing alternative alleles at a marker located very close to a QTL.

In practice neither the map positions of the markers nor their means are known with very great accuracy because the number of DH lines and replicated plots are generally few and the heritability of quantitative traits is generally low. Moreover, linked QTLs create difficulties because their individual effects can combine either to create the impression of an intermediate, ghost QTL, or to conceal their effects altogether.

Various analytical procedures have been developed to maximize the efficiency and accuracy of locating individual QTLs and to separate linked QTLs. They all rely on the relationships described in the previous paragraph and use either maximum likelihood or weighted least squares regression procedures to estimate the QTL locations and effects that best fit the observed trait scores for each genotype.

The main difficulty is that there are two unknown parameters with respect to each marker score; the QTL effect, a, and map position. This means that it is necessary, in all methods, to use an iterative approach which involves trying all possible QTL locations along each chromosome and identifying the position or positions of the QTL which best fit the observed data as indicated by the size of the likelihood or the residual regression variance. As a result, a very large number of statistical tests are performed with the concomitant risk of false positive results.

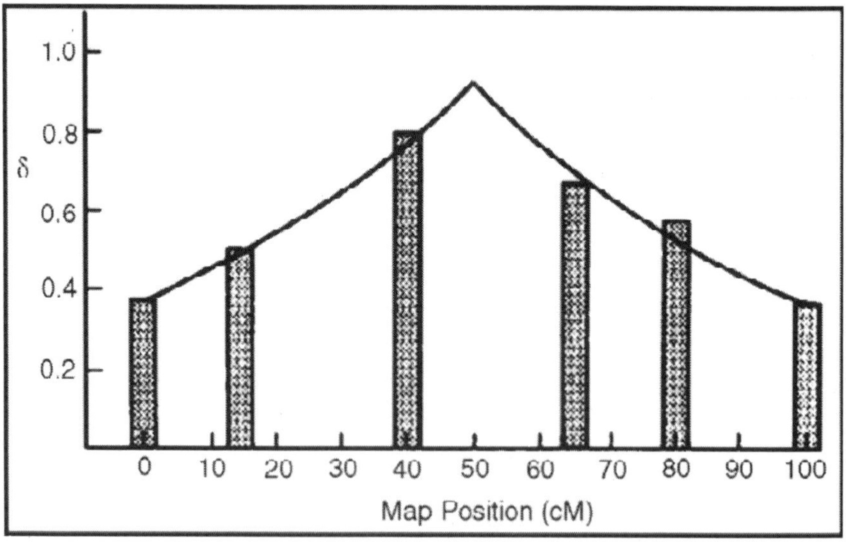

Fig. Quantitative Trait Effects (g) Associated with Marker Loci.The Bars Indicate the Observed Effects,While the Curve Illustrates the Expected Relationship if a QTL of g = 1.0 Existed at 50 cM

All methods provide estimates of QTL locations and effects with similar precision in terms of confidence intervals, and this precision drops rapidly as the heritability of individual QTLs declines, i.e., the estimates are accurate only when there are just two or three unlinked QTLs with large effects. For example, Hyne et al. (1995) showed that the 95per cent confidence interval for the location of a QTL contributing 10per cent to the phenotypic variance of an F2 could be as large as 35 cM. However, conservationists and breeders who are concerned with introducing QTL into commercial cultivars from other cultivars or more distantly related species, would normally only be interested in those cases where a few QTLs of major eff ect are to be manipulated. There is no evidence that dense maps are required for initial

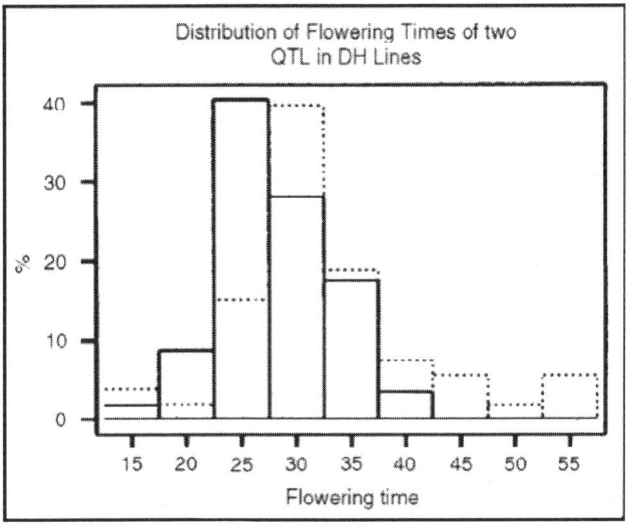

Fig. Distribution of Flowering Times Among Doubled Haploid (DH)Lines in Brassica Oleracea Containing alrernative Alleles of a Marker Linked to a QLT.The two sets Significantly Different Means Indicating a QLT with an Effect of Apporoximately ± 2 Days

QTL location. It is better to have large populations and a few (5–10) markers per chromosome.

There is evidence that many of the larger QTLs located so far map closely to previously known major genes. This is often used as evidence that there are, in fact, very few QTLs, but this may well be misleading for several reasons. The QTL map locations are sufficiently imprecise that they could well appear spuriously close to candidate loci. Only QTLs of large effect will stand much chance of being located and, by definition, they are likely to be major genes. It is likely that many cases where QTLs have very large effects could be due to the chance association of alleles of like effect at several QTLs along a chromosome, which cannot be separated as individual effects.

Much effort has been devoted to increasing the precision of QTL location, in particular in improving the power of the test to reduce the failure to de te ct ge n u in e Q TL . A l on g- est ab lis he d probl em w ith b re e di n g f or quantitative traits in plants is the genotype 3 environment interaction. Using QTL location techniques, it is now possible to explore these effects at the level of individual QTLs although it is often difficult to know whether genotype 3 environment or poor repeatability is responsible.

ALIEN GENE TRANSFER

Once useful genes have been located on a framework map, it is then possible to use the molecular markers to facilitate the transfer of the useful genes between species and strains in an efficient manner. As stated earlier, recombination is not a frequent event along a chromosome and, furthermore, two recombination events rarely occur within a distance of 15 to 20 cM of each other.

It therefore follows that if a useful target gene (a major gene or a QTL) is known to be located within a region of chromosome flanked by two markers that are no more than 15 cM apart, then these markers can be used to follow and control the progression of the target gene through successive stages of a breeding programme. Clearly, the alleles at the marker loci have to differ in the donor and recipient cultivars, but any chromosome that contains A1 and B1 at the marker loci will also contain T1, the target allele to be transferred.

Should single recombination occur close to T, then the chromosome will contain A1 or B1 but not both. Providing the markers are less than 15 cM apart double recombination resulting in A1T2B1, and hence, loss of T1, can be safely ignored. If the location of T is not accurately known, then three or more marker loci may be needed to be sure of safely bracketing the region containing T without fear of double recombination.

There are conflicting requirements in marker-assisted gene transfer: the need to keep the bracketed region large enough to be sure of holding the gene to be transferred whilst not having it so large that too many other linked but undesirable alleles are transferred with it. Let us now consider the basic procedure of gene transfer.

Traditionally, breeders have introduced a useful allele into their cultivars by backcrossing the original F1 to the commercial cultivar whilst selecting at each generation for the desired allele. As was shown earlier, this can result in a very large region of chromosome around the target gene surviving even after ten generations of backcrossing – a phenomenon known as linkage drag. With markers, however, it is possible to reduce linkage drag considerably whilst simultaneously reducing the number of backcrossing generations to two or three instead of the normal six to eight. Moreover, the amount of plant material raised at each generation can be reduced because much of the molecular genotyping can be achieved at the juvenile stage.

The aim is to hold the target allele, T1, heterozygous throughout the backc rossi ng proc ess, by s elec ti ng fo r th e fla nkin g marke r s, w h ils t simultaneously selecting for the genotype of the recurrent, commercial cultivar at all other loci. The latter can readily be achieved by having as few as four or five well-spaced markers on all chromosomes and selecting for recurrent parent alleles at each generation.

Depending on the chromosome number, some individuals in the first backcross generation will be homozygous for several whole chromosomes whilst still being heterozygous for the target gene. Those which have most of the recurrent parent chromosomes are selected and backcrossed again, and if necessary the process is repeated for a third generation. For each genera tion, fewer markers need to be scre ened as those found to be homozygous in the previous backcross no longer have to be checked.

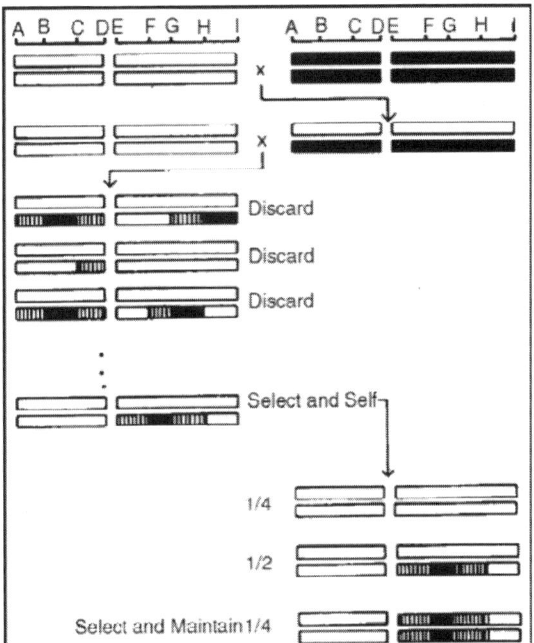

Fig.The use of Molecular Markers to Facilitate Gene Introgression.

Unless the position of the target gene is very accurately defined, it is prudent to have several markers covering its likely position, again with the proviso that they need to be separated by no more than 15 cM. At the final s tage of th e i ntr og re ssion pr oce ss, a ba ckc ros s in dividu al wh ich is homozygous for most of the recurrent parent alleles is selfed or, if this is not possible, intercrossed with a similar genotype. From among the progeny, individuals are chosen which have different combinations of markers bracketing the target region as shown in figure.

These can be selfed again and homozygotes for the different sequences identified and multiplied for reassessment for the target trait. Some will fail to show the target trait because of recombination, but of those that do, the lines with the shortest sequence of donor markers are selected for further tr ia ls . The approa ch c an ob vi ously be adapted to the s imu lt aneo us introgression of several genes: no new principles are involved but the screening process becomes more elaborate. It is always possible at a later stage to reduce the length of the introgressed region around T by looking at other, more closely linked, markers. Similarly, it is wise to check the origins of the other chromosomes to avoid having introgressed unwanted regions through missing double recombinants in the central regions of individual chromosomes or single recombinants towards their ends.

If such errors are found they can easily be corrected by another round of backcrossing to the recurrent parent. When attempting to reduce the length of donor chromosome surrounding the target gene, it is important to realise that this should be done in two successive generations. If the sites of the re quire d re c om b ination ev e n ts are le ss t h an 1 5 c M a pa rt, th en the simultaneous double event will not occur.

Even if they are far enough apart for the double event to occur, the probability of such an event may be too low to be practicable. If two recombination events each have a probability of 0.05 (i.e., 5per cent), their combined probability is at best 0.0025. Therefore, it would be necessary to genotype approximately 1200 individuals to be 95per cent sure of having at least one double recombinant. If it was tackled in two stages then it would require only the genotyping of 86 (= 28 in the first round plus 58 in the second; less being needed in the first round because the recombination could occur on either side of T), a very considerable saving in effort of 93per cent, or 1114 plants!

Marker-aided selection has been successfully tried for quantitative traits in several crops while the efficiency of the procedure has been examined by Lande and Thompson (1990).

MAP-BASED GENE CLONING

Providing a target gene can be located to within a pair of marker loci, it should be possible to use standard cloning techniques to walk along the chromosome between the markers and to find the target gene. The feasibility of this depends on the distance between the markers and the ease with which the target gene can be recognized from among the clones.

Depending on the species, a genetic length of 1 cM can, on average, consist of as few as 280 kbp of DNA in rice, or 2220 kbp in maize. It is not always easy to locate a target gene with sufficient accuracy to be able to say that it is between two markers as close as 1 cM, and hence, the actual lengths of DNA between them can often be much greater. With QTLs one can seldom achieve anything approa ch in g this accura cy by ex is ting meth ods of conventional mapping. Furth ermore, these are average distances per centimorgan; some chromosomal regions could be much longer or, indeed, less.

Increases in mapping reliability can be obtained by using what are called near isogenic lines (NILs), which are lines obtained by selfing or backcrossing and are known to differ from some standard genotype by just a short defined section of chromosome. Providing this section is delineated by accurately mapped flanking markers, and the NILs which do and do not contain this region can clearly be shown to differ for the target gene, then the target gene has to be in that region. The genetic and physical proximity of the markers can be established by conventional mapping and by gene cloning.

Map-based gene cloning involves starting from one of the markers and walking to the next by means of a series of partially overlapping cosmid clones. Cosmid clones are necessary because they can be up to 50 kb in length and hence, require fewer cloning steps in order to cover the DNA between the markers (280 to 2200 kb per cM in the examples cited above). Identifying the target gene is more difficult. Clearly, if the gene is between the two flanking markers it has to be on one of the cloned regions and could possibly be on two overlapping clones.

It could be identified by transforming plants with each clone separately and identifying which transformant expresses the effect. If the gene concerned ex hibits c la ssica l domina n ce it would be necess ary to tran sform the homozygous recessive with the dominant allele or the transformant would not be recognized. Alternatively, the gene has to be introduced into a species that does not normally express the gene. Once the effective cloned fragment is identified, the actual location of the gene within the fragment could be identified by successively transforming sub-fractions of the clone or, following sequencing, by looking for potential open reading frames. In those cases where the target gene produces a known product, cDNA clones derived from tissue likely to be rich in the appropriate mRNA of the target gene can be used to identify likely sequences in the previous overlapping clones.

Once the gene has been identified by one of these methods, its structure and activity can be studied in detail. Disease resistance genes are obvious candidates for such studies as their action is very specific and their location on the genome of many species is well documented. However, unless suitable candidate loci.

❐

6

Plant Growth in Different Ways

New cells form continually in the apical meristems. Cells enlarge slowly in the apical meristem and more rapidly in the subapical regions. The resulting increase in cell volume can range from severalfold to one hundredfold, depending on the species and environmental conditions. Classically, plant growth has been analysed in terms of cell number or overall size (i.e., mass). However, these measures tell only part of the story.

Tissue growth is neither uniform nor random. The derivatives of the apical meristems expand in predictable and site-specific ways, and the expansion patterns in these subapical regions largely determine the size and shape of the primary plant body. The total growth of the plant can be thought of as the sum of the local patterns of cell expansion.

The analysis of the motions of cells or "tissue elements" (and the related problem of cell expansion) is called kinematics. In this section we will discuss both the classical definitions of growth and the more modern, kinematic approach. The advantage of the kinematic approach is that it allows one to describe the growth patterns of organs mathematically in terms of the expansion patterns of their component cells.

Growth in plants is defined as an irreversible increase in volume. The largest component of plant growth is cell expansion driven by turgor pressure. During this process, cells increase in volume manyfold and become highly vacuolate. However, size is only one criterion that may be used to measure growth.

Growth also can be measured in terms of change in fresh weight—that is, the weight of the living tissue—over a particular period of time. However, the fresh weight of plants growing in soil fluctuates in response to changes in water status, so this criterion may be a poor indicator of actual growth. In these situations, measurements of dry weight are often more appropriate. Cell number is a common and convenient parameter by which to measure the growth of unicellular organisms, such as the green alga Chlamydomonas. In multicellular plants, however, cell number can be a misleading growth measurement because cells can divide without increasing in volume.

Growth is assessed by a count of the number of cells per milliliter at inc reas ing times a fter the c ells a re pla ced in fres h growth medium. Temperature, light, and nutrients provided are optimal for growth. An initial lag period during which cells may synthesize enzymes required for rapid growth is followed by a period in which cell number increases exponentially.

This period of rapid growth is followed by a period of slowing growth in which the cell number increases linearly. Then comes the stationary phase, in which the cell number remains constant or even declines as nutrients are exhausted from the medium.

For example, during the early stages of embryogenesis, the zygote subdivides into progressively smaller cells with no net increase in the size of the embryo.

Only after it reaches the eight-cell stage does the increase in volume begin to mirror the increase in cell number. Because the zygote is an especially large cell, this lack of correspondence between an increase in cell number and growth may be unusual, but it points out the potential problem in equating an increase in cell number with growth.

Although cell number may not always be a reliable measure of plant growth, under most circumstances dividing cells, particularly in meristems, double in volume during their cell cycle. Therefore, an increase in cell number, such as the increase brought about by the activity of the apical meristems, does contribute to plant growth. However, the largest component of plant growth is the rapid cell expansion that occurs in the subapical region after cell division ceases.

Because all the cells of the plant axis elongate under normal conditions, the greater the number of cells produced by the apical meristem, the longer the axis will be. For example, when Arabidopsis plants are transformed with a gene that encodes cyclin, a key component of the cell cycle regulatory machinery, the cells of the apical meristem progress through their cell cycles more rapidly, so more cells form per unit time.

As a result, the roots of these transgenic plants have more cells and are substantially longer than the roots of wild-type plants grown under similar conditions.

New cells form continually in the apical meristems. With each new round of cell division and associated cell expansion, the older derivatives are displaced a small distance from the apex. As the cells recede farther from the apex, the rate of displacement is greatly accelerated. By viewing plant growth as a process of cell displacement from the apex, we can apply the principles of kinematics.

PRIMARY GROWTH OF PLANT

The primary growth of a plant body is brought about by the activity of primary meristems, particularly by apical meristem.

Details about primary growth occurring in stems, leaves and roots, may be studied by observing a transverse structure is differentiated into outer epidermis, the inner ground tissue and vascular tissue.

- Dicotyledonous stem: Example: Sunflower (Helianthus annus). The transverse section is almost circular in outline.
- Epidermis: This is the outermost layer, uniseriate, covered with thin cuticle, and bears a few stomata and many multicellular trichomes.
- Ground Tissue: This fundamental tissue is differentiated into outer cortex and central pith.
- Cortex: This is many layered and differentiated into hypodermis and inner cortex. Hypodermis is found below the epidermis, and is constituted with angular collenchyma which is 3-5 layered

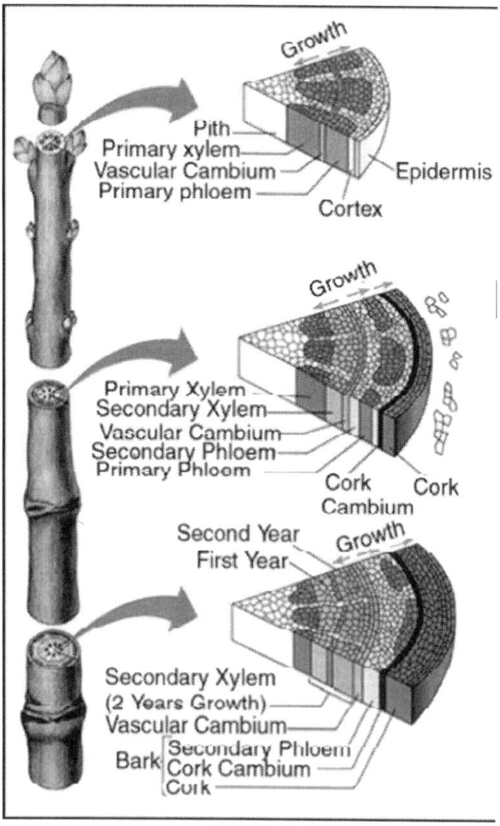

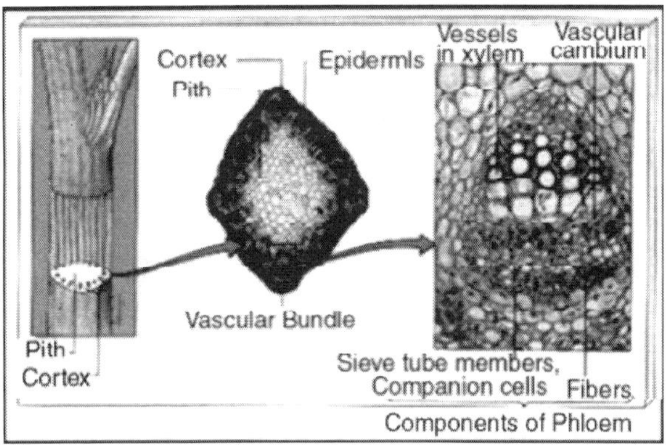

The inner part of cortex is constituted with chlorenchyma and parenchyma; large numbers of intercellular spaces are present between these cells. The inner most cortical layer is wavy in outline, rich in sta rch grains, with barrel shaped cells and is called endodermoid layer. (Endodermis is generally absent in stem; endodermoid layer differs from endodermis in lacking casparian thickenings).

- Pericycle: This is the outermost region of vascular tissue. This consists of alternately arranged patches of sclerenchyma and parenchyma. The sclerenchyma is also known as hard bast fibres. Since, it is found externa l to vascular bundle, the patch of sclerenchyma is also called bundle cap.
- Vascular tissue: This consists of many vascular bundles arranged regularly in the form of broken ring (Eustele). In each vascular bundle the xylem and phloem elements are in the same line; the presence of one to few stripped intrafascicular cambium is found in between xylem and phloem; the cambial cells are thin walled and are almost rectangular phloem is only external to xylem: the protoxylem is nearest to centre. The vascular bundles are conjoint, collateral, endarch, open and wedge shaped.
- Pith or Medulla: This is inner to vascular tissue and occupies the cen tral region. It is composed of cells of paren chyma, w ith intercellular spaces. Extensions of these cells into the areas in between vascular bundles are called medullary rays.
- Monocotyledonous stem: Example: Zea mays (Maize). The outline is circular.

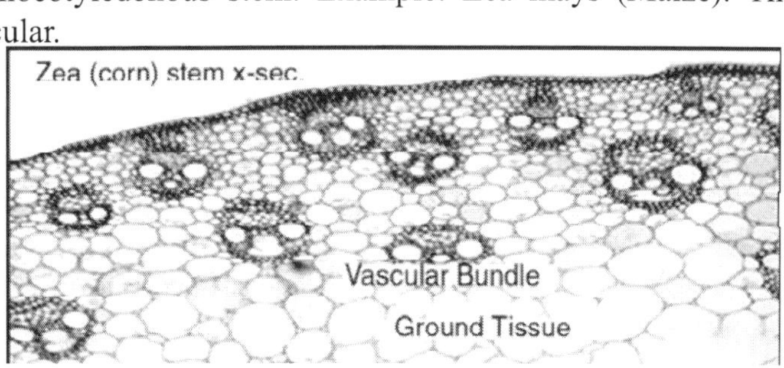

- Epidermis: This is the uniseriate, outermost layer and is covered with thick cuticle. Trichomes are absent.
- Ground tissue: The ground tissue is not differentiated; into cortex and pith, since, there is scattered arrangement of vascular bundles. Even endodermoid layer and pericycle are also not differentiated. Hypodermis constitutes the exterior ground tissue, and is composed of 2-3 layers of sclerenchyma. The major part of the ground tissue is parenchymatous.
- Vascular tissues: This consists of numerous, oval shaped vascular bundles sca ttered throughout th e ground tissue, hence, th e arrangement of vascular tissue is called atactostele. Peripheral vascular bundles are small-sized but many in numbers. The central vascular bundles are large sized but a few in numbers in all the vascular bundles the xylem and phloem are in the same line; cambium is absent; phloem is only external to xylem; protoxylem faces the centre.

Hence, the vascular bundles are described as conjoint, collateral, endarch and closed. Each vascular bundle is surrounded by a bundle sheath of sclerenchyma, which is prominent at the base and tip of vascular bundle.

Xylem comprises usually 2 oppositely arranged Metaxylem elements (pitted, facing periphery) and a few protoxylem elements (facing the centre of stem) in a linear row.

If an imaginary line is drawn connecting the two Metaxylem elements with that of protoxylem, xylem appears 'Y' shaped.

Usually the lower protoxylem elements breakdown and form a lysogenous cavity. Above the metaphloem, a small band of obliterated protophloem is seen. Metaphloem consists of sieve tubes and companion cells but not phloem parenchyma.

Comparison between the stems of Dicot and Monocot:

		Dicot stem Example: Sunflower	**Monocot stem Example: Maize**
1.	Epidermal trichomes	Present multicellular	Absent
2.	Hypodermis	Collenchymatous	Sclerenchymatous
3.	Ground tissue	Differentiated into cortex and Pith	Undifferentiated
4.	Endodermoid layer and Peri-cycle	Present	Absent
5.	Stele	Eustele	Atactostele
6.	Vascular bundles	Many open, wedge-shaped	Numerous, closed, oval-shaped
7.	Lysigenous cavity	Absent	Present
8.	Bundle cap	Present	Absent
9.	Bundle sheath	Absent	Present
10.	Secondary Growth	Occurs later	Does not occur

Dicotyledonous root: Example: Lablab purpureus (Beans)

- ***Epiblema:*** The external protective layer is called Epiblema or Piliferous layer (the term epidermis is generally not applied to roots). It is uniseriate. Cells are thin walled. Some of these cells extend into unicellular root hairs.
- ***Cortex:*** The region inner to epiblema is called cortex and it is homogenous. The cortex is composed of thin walled, spaciously arranged parenchyma cells that constitute many layers.

With increase in age, the exterior cortical region becomes suberized and dead. This exterior cortex is known as exodermis.

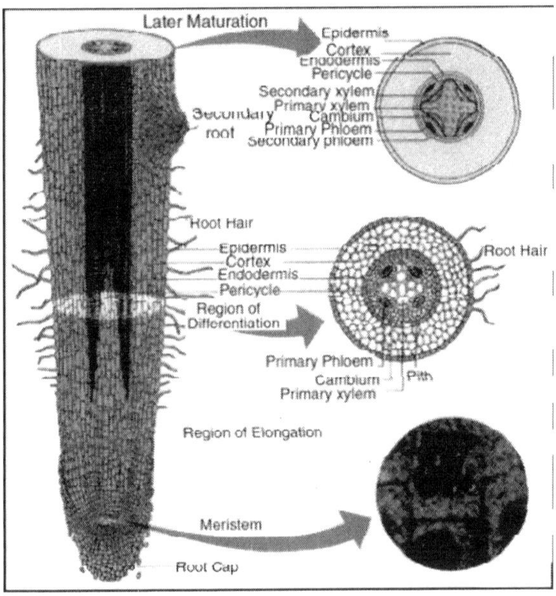

The innermost cortical layer is called endodermis. Its cells are barrel- shaped and are closely packed. Many cells get suberized on their radial and inner walls. These thickenings are called casparian stripes. Such endodermal cells may become totally suberized on all wall surfaces and are called stone cells.

These cells do not allow conduction. The non-suberized endodermal cells are called passage cells. Inner to endodermis, one or two layers of parenchyma are present, constituting pericycle.

- ***Stele:*** Xylem and phloem patches are in equal number and alternate with each other. Xylem or phloem patches are usually four in number. In Xylem, the protoxylem elements face pericycle; between xylem an d phloem e lements , pa rench yma cel ls are presen t constituting conjunctive tissue, which later becomes a secondary meristem.

Hence, vascular bundles are described as radial, exarch and open.

Pith is very small, with closely packed parenchyma cells.

Monocotyledonous root: Example: Canna

Outline almost circular

The nature of epiblema, exodermis, cortex, endodermis and pericycle is similar to that of Dicot root. However in cortex, large air spaces of schizogenous origin may be seen. The nature and arrangement of vascular bundles are also similar to that of Dicot root. But vascular bundles are usually more than six in number. The Parenchyma present in between xylem and phlegm is called conjunctive tissue, with thin cell walls. This tissue does not become meristematic and thus differs from Dicot root. Pith is relatively large and parenchymatous.

Pith and conjunctive tissue, become lignified and Sclerenchymatous. The vascular bundles are described as polyarch, radial, exarch and closed. An endogenous lateral branch may be seen developing from pericycle and passing through cortex.

The leaf is horizontal and dorsiventrally differentiated (upper = ventral, lower = dorsal).

- ***Epidermis:*** The two surfaces are bounded by their respective epidermal layers. The epidermal cells are barrel-shaped, compactly arranged; upper epidermis is covered with thick cuticle and lacks stomata; lower epidermis is light green, covered with thin cuticle and is interrupted by stomata.
- ***Mesophyll:*** This is the ground tissue. It is differentiated into upper palisade and lower spongy parenchyma. Palisade parenchyma is found immediately below the upper epidermis, 2 to 3 layered, with compactly arranged tubular cells, rich in parietal chloroplasts.
- Spongy parenchyma is found above the lower epidermis; these cells are varied in shape and sizes, very loosely arranged enclosing air spaces some of which open into stomata. Chloroplasts are parietal in the parenchyma cells.
- ***Vascular tissue:*** Vascular bundles vary in size; each bundle is conjoint, collateral and closed. The vascular bundle is covered by a bundle sheath of parenchyma cells. An extension of bundle sheath is seen above and below the vascular bundle, reaching the epidermal layers. Phloem is towards lower epidermis; xylem is towards upper epidermis, with metaxylem facing phloem. Fibres are absent in both xylem and phloem. The xylem and pholem elements are conspicuous only in large vascular bundles.

Monocotyledonous leaf Example: Maize

The leaf is Isobilateral in nature and is vertically oriented.

- Epidermis: This is uniseriate, with barrel-shaped, compactly arranged cells and is covered with thick cuticle.

  Stomata are found on both upper and lower epidermal layers hence, it is called amphistomatic (more on the lower epidermis). Though the leaf is referred to as isobilateral, it is only in upper epidermis, a few large, empty and colourless bulliform or motor cells are present. During dry weather, these motor cells help the leaf to role over, due to the changes in turgidity. This rolling of leaf reduces the rate of stomatal transpiration.

- Mesophyll: There is no differentiation of mesophyll into spongy and palisade parenchyma. All the cells of chlorenchyma are alike, isodiametric, almost compactly arranged with numerous parietal chloroplasts.
- Vascular tissue: The vascular bundles are numerous, arranged in parallel series (venation is palmate-parallel), conjoint, collateral and closed. Phloem is towards lower epidermis. Each vascular bundle is surrounded by chlorenchymatous bundle; this sheath also serves for temporary storage of starch. A few vascular bundles are larger in size, with more amount of xylem and phloem and with large bundle sheath cells. A patch of sclerenchyma is present above and below the large sized vascular bundles.

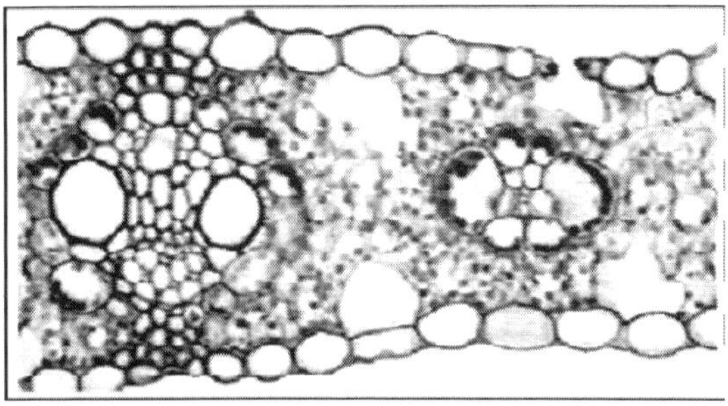

Table. Comparison Between the Leaf Structures of Dicot and Monocot.

	Dicot Root **Ex: Mango**	**Monocot Root** **Ex: Maize**
1. Orientation of leaf	Horizontal, dorsiventarlly differentiated	Vertical Isobilateral
2. Stomata	Usually on lower epidermis	On both epidermal layers
3. Cuticle	Thick on upper epidermis	Thick on both epidermal layers
4. Motor cells	Absent	Present in upper epidermis
5. Mesophyll	Differentiated into upper palisade parenchyma and lower spongy parenchyma	No differentiation of mesophyll
6. Vascular Bundles	All of them are not seen in parallel series and are supported by bundle sheath extension	Vascular Bundles are in parallel series and are supported by sclerenchyma patches

PLANT GROWTH INTERACTIONS

Soil is a medium for plant growth, and its structure should not hinder the movement of water, oxygen and nutrients to plant roots, impede the growth of roots, or allow the buildup of toxic substances around roots. Soil structure can be described as the organization of the particles in the soil— the internal configuration of the soil matrix.

Soil structure is formed by many interacting processes in the soil, especially shrinkage with drying, channel formation by plant roots, faunal activity, and cultivation.

Well-developed soil structure is represented by distinct stable aggregations (peds) of soil particles separated by interped pores. The interped pores of an aggregated soil are larger, on average, than pores between particles of unaggregated soil, resulting in a less dense condition.

The larger pores are also critical in terms of allowing movement of air and water through the soil. Unfavourable soil structure, on the other hand, can impart the characteristics of greater soil density, decreased aeration, poor percolation, and mechanical impedance to roots, which can restrict plant growth. There are a number of factors that may result in poor soil structure. Soil texture is important; in fact, very sandy soils are usually classified as structureless due to lack of soil matrix (sand is viewed more as the "skeleton" of the soil). In sandy soils, size distribution and packing of the grains are considered rather than structure per se.

In loamy and finer-textured soil, clay acts as a cementing agent to provide some level of stability to the particle orientation in structured soil; however, clay can also cement particles in a structurally degraded condition ("massive"). Chemical conditions can be responsible for structurally degraded soil; a relatively high sodium content causes dispersion of particles and elimination of aggregation.

These "sodic soils" occur naturally but can also be initiated or exacerbated by human activity. The human-imposed factor that degrades soil structure to the greatest extent is compaction. This chapter focuses on the effects of compacted soil on plant growth, including the direct effect of mechanical impedance as well as the indirect effects that usually accompany compaction. Finally, amelioration of compaction is discussed, with emphasis on the ability of plant roots to effect changes in soil structure and hence, modify physical properties.

COMPACTION

Hillel described soil compaction as the compression and densification of an unsaturated soil body, reducing the fractional air volume. The degree to which a soil will compress is a function of the soil moisture content, antecedent bulk density, magnitude of the compactive effort, soil texture, and content of organic matter. Forces that compact soil can originate from natural sources, including rainfall, radial root pressure, cycles of wetting/ drying and freezing/thawing, or traffic of people and machinery. The potential to degrade soil structure has increased as modern machinery has become larger and heavier, causing compaction to a greater depth. Compounding this problem is the tendency to tread repeatedly over soil while implementing cultural practices such as seeding, fertilization, pest control, harvesting/ mowing, or construction during site development.

Soil compaction on recreational sites is thought to be one of the most challenging issues faced by turfgrass managers. Cultivation implements are generally designed to overcome the bonds stabilizing structure, breaking aggregates into smaller ones for seedbed preparation to maximize seed-soil contact, enhance water infiltration and retention characteristics, and improve soil aeration. Routine tillage of soil, however, can also degrade soil structure by sh atte rin g a ggr ega tes , elimi n ating por e n etw orks , an d imp osi n g compacting forces, particularly in a subsurface layer of soil.

Soil Responses

Porosity, bulk density, and water retention characteristics are physical properties cha rac terizin g soil structure and compaction phenomena. Compaction of a soil decreases soil porosity and subsequently increases soil bulk density.

Veihmeyer and Hendrickson showed that the critical soil bulk density needed to inhibit sunflower (Helianthus annus L.) root growth varied with texture; no roots penetrated soil with a bulk density of 1.9 g cm–3. The lowest density where roots failed to penetrate was 1.46 g cm–3 for an Aiken clay loam. Air tends to occupy the larger pores of soil, whereas the affinity of water for mineral surfaces creates water films and wedges (menisci) within the small fissures and capillaries of the soil. A reduction in the total porosity of soil, as a result of compaction, generally occurs at the expense of the air porosity, those relatively large pores not filled with water at a given soil water potential (usually –33 kPa to represent "field capacity").

Swartz and Kardos observed a decrease in air porosity from 19.3 to 12.7%, with increasing levels of compaction on several sand-soil-peat mixes differing in moisture content at the time of compaction, whereas capillary porosity increased from 32.2% at the lowest compaction level to 35.7% at the highest level. In aggregated soils, the loss of porosity due to compaction can be attributed to the collapse of interaggregate pores. Water retention and flow characteristics in soil are influenced by alterations of soil porosity. Relative to noncompacted soil, compacted soil will generally retain a greater amount of water at a given soil matric potential as a result of increased capillary porosity. Soil water permeability, particularly at saturated or nearly saturated conditions, is a function of air porosity; as air porosity decreases, so does water permeability. Compacted soils with lower infiltration rates are prone to problems with stagnant water and anaerobic conditions.

CONSTRAINTS ON ROOT GROWTH

Soil Aeration

Soil oxygen movement to plant roots is critical to maintain adequate respiration for growth. Anaerobiosis, or oxygen stress, occurs in soil when the rate of supply falls below the biological demand. Impeded soil aeration can arise from poor drainage and waterlogging or from compacted soil. One approach to evaluate soil aeration is to measure the composition of soil air. This technique indicates that gas exchange between the atmosphere and soil air is restricted when oxygen content of the soil air falls significantly below that of the atmosphere. An important challenge with this approach is how to extract a representative sample of soil air that avoids mixing of gases from outside the point of sampling or contamination from the atmosphere.

Another approach for assessing soil aeration is to determine the air porosity, or fractional air space, at a standard soil water potential. Work to identify threshold air porosity values below which plant growth is limited has generated values ranging from 5 to 20%. Madison has indicated that 10% air porosity is the threshold level for intensively utilized turf. The United States Golf Association Green Section (USGA) has suggested 15% air porosity at a 3 kPa water tension as the lower test limit for materials used in the construction of root zones for golf course putting greens. The extent to which air porosity in the root zone changes with time under a perennial turf system is a subject of current studies.

The rate of air exchange, rather than the volume of soil air, is the more important factor in soil aeration and therefore limits the usefulness of air porosity as an index of soil aeration. Gaseous movement in the soil occurs by two main processes: convection and diffusion. Convection, where the moving force is a gradient of total gas pressure, is thought to be a minor mechanism of soil aeration except at shallow depths and in soils with large pores. Although convection is not considered an important factor for more deeply rooted plant species, it could be more influential for shallow-rooted turfs. A large fraction of the root system of moderately to highly maintained turfs is found at surface depths of 5 cm or less.

Furthermore, the crown, the major meristematic organ, of turfgrass plants is located near the soil surface. Thus, turfs with a high plant population could have sufficiently high oxygen demands at the soil surface that convection would play a useful role in maintaining adequate aeration for plant crowns and surface rooting. Convection, therefore, could be partially responsible for the observations of improved turfgrass quality after changes in air masses (pressure) associated with passing weather systems. Below the surface, however, diffusion is considered the more important exchange mechanism in soil, and the moving force is a gradient of partial pressure. Gas exchange between the atmosphere and soil is maintained predominantly by diffusion through air-filled pores. In contrast, live tissue is typically hydrated, and the supply of oxygen to roots occurs by diffusion through water films. This is an important distinction, since, the diffusivity of oxygen in water is less than its value in air by a factor of 10"4.

The solubility of oxygen in water may be a compounding issue, especially at higher temperatures, where solubility decreases. Methods that measure gaseous diffusion through air-filled pores do not provide information on the impedance to oxygen presented by the water film surrounding a root. It is possible that an inadequate supply of oxygen to roots can occur in a well- aerated soil (i.e., high air-filled porosity and O2 concentration) if the roots are surrounded by thick water films. Thus, under such conditions, measurement of soil oxygen movement to plant roots would provide a more complete assessment of soil aeration. Lemon and Erickson introduced the method commonly called ODR (oxygen diffusion rate) to measure oxygen diffusion to a root-like electrode inserted into the soil and enveloped in water films of the soil. The ODR technique has been used by a number of researchers to demonstrate the reduction in oxygen supply in compacted soil.

Limiting soil aeration conditions are often transient and strongly affected by wetting and drying cycles in compacted soil. Therefore, the timing of gas composition and ODR measurements is critical for accurate detection of anaerobiosis. Generally, 200 ng cm–2 min–1 has been considered the minimum ODR value, below which plant roots will not grow. Some researchers have reported values ranging from 50 to 200 ng cm–2 min–1 as limiting ODR values for turfgrass root growth. The interpretation of these results has been questioned because, in addition to the diffusion of oxygen, other soil factors and components could have affected the measured ODR.

Work by Blackwell has helped to reduce errors associated with reactions other than the reduction of oxygen in the ODR technique. Moreover, early studies of limiting ODR were performed in growth chambers at about 20æ% C, whereas soil temperatures in the field can vary considerably during the growing season. Blackwell and Wells concluded that the effect of oxygen flux on root growth is related to temperature differences and differences in root respiration. Thus, it is plausible that the oxygen flux limiting root growth under high soil temperatures may be greater than values that have been observed at 20°C.

When oxygen becomes limiting, some species of bacteria shift metabolic pathways so as to utilize other compounds as terminal electron acceptors. Thus, plants may have not only to survive periods without oxygen but also to withstand toxic substances such as hydrogen sulfide, which are produced during anaerobic microbial respiration. Development of aerenchyma tissue (intercellular spaces) that allows diffusion of oxygen from aerial plant tissue down to roots may be an adaptive response that helps certain species of plants to survive periods of soil anaerobiosis. Agnew and Carrow reported increased root porosity in Kentucky bluegrass (Poa pratensis L.) grown under compacted soil conditions. They observed the greatest root porosity under conditions of both compaction and water stress.

Soil Strength

Soil strength can be described as the capacity of soil to resist a force without rupture, fragmentation, or flow. Various methods exist to quantitate soil strength in the laboratory, but the method most used in the field is the penetrometer. It measures the effort necessary to push a thin, cone-tipped probe into the soil. Taylor and colleagues conducted pioneering research demon strating that th e strength of a s oil, a s mea sured with a static penetrometer, will increase as soil bulk density increases and water content decreases; consequently, mechanical impedance to root growth is greater. The limiting strength of soil to cotton (Gossypium hirsutum L.) root growth is in the range 2.5 to 3 MPa. For many crops, a 2-MPa penetrometer resistance is commonly encountered under field conditions and can reduce root length and elongation by at least 50%.

Root Morphology

In experiments designed to study the constricted growth of roots, the ability of ryegrass (Lolium perenne L.) roots to enter rigid pores (glass capillaries and sheets of steel) depended on the size of the root cap and the thickness of the stele, which needed to be about one-third of the nominal root thickness. Roots could elongate down long capillaries while constricted, albeit at a reduced rate. Roots can enter pores of diameter smaller than the root tip itself only if the rigidity of the pore structure is weak enough to allow soil displacement. Thus, major root axes typically thicken when soil pores are too small for root penetration. Ethylene production has been linked to this thickening response of roots subjected to mechanical impedance. Abscisic acid, which increases temporarily under conditions of mechanical impedance, has also been shown to induce root thickening as well as increased root-hair number and curling of roots. The thickening of the root compresses soil laterally and minimizes friction with soil as the root extends axially. Additionally, root-hair development aids in anchoring the root as it penetrates compacted soil. Thus these responses can be viewed as adaptive, as they mitigate the effects of axial resistance on growth. Additional contributing factors to thicker roots' relative endurance of compact soil might be greater resistance of thicker roots to bending or higher axial pressures exerted by thick roots. Variation in plant genetics exists for tolerance to high mechanical impedance. Interspecific comparisons have shown differences in root dimensions and the ability to penetrate soils. Greater root thickness and the tendency for roots to expand radially in response to mechanical impedance are correlated with the capacity to elongate in hard soil. Intraspecific studies have also demonstrated genetic variation in characteristics associated with tolerance to mechanical impedance. Two tall fescue (Festuca arundinacea Schreb.) lines with large-diameter roots were able to penetrate a hardpan at the 0.4 to 0.6 m soil depth, extracted more soil water from the 0.6 to 1.2 m depth, and yielded 40% more in dry matter compared to fescues with small- diameter roots.

Development of efficient techniques to screen plant germplasm for greater tolerance to high mechanical impedance could prove to be highly useful. Iijima et al. described the morphological development of rice (Oryza sativa L.) and maize (Zea mays L.) root systems as affected by soil compaction. Elongation of the main root axes of rice and maize was restricted in compacted soil. Seminal roots of maize were much less restricted than those of rice. The root system of rice was characterized by a larger fraction of long, thick laterals with potential to produce higher-order laterals on their axes. Growth of higher (second- and third-) order lateral roots in compacted soil compensated for the restricted growth of the main root axes in both species. The work of Carrow and colleagues indicates that the effects of compacted soil on root growth varies with plant species and management. Total root mass of Kentucky bluegrass decreased in response to compaction treatment, whereas tall fescue and perennial ryegrass root systems were more tolerant to compaction stress.

Subsequent work indicated that the most detrimental effect of compaction on root growth of tall fescue and perennial ryegrass occurred under higher nitrogen fertilization. Although total root mass of a turf may not change under compacted conditions, a redistribution of the root system can occur where root mass increases in the surface 5 cm of the soil at the expense of rooting below the 10-cm depth. Root mass may not provide the most detailed index of a root system's response to compacted soil, since, root mass is affected less by bulk density than root length is, primarily because of larger-diameter roots in high-density soil. Thus, the measurement of the number and length of roots growing in soil may be a more sensitive indicator of compaction stress on a root system than root mass. Root length and number measurements, however, are expensive, particularly for experiments evaluating a large number of treatments and for plant systems such as turf, which have extremely high root lengths.

Physical alteration of water and nutrient uptake by a root system can result as the orientation and morphology of roots change. Higher soil bulk density brings more soil, and associated water and nutrients, into contact with the surface area of the root system and thus has the potential to increase availability. This effect is particularly pronounced for nutrients such as phosphorus, which have relatively low mobility in the soil. Greater soil contact with the root system, and thus availability of nutrients and water, could at least partly explain the greater clipping yields of turf growing on heavily compacted plots compared to lightly compacted plots during the spring. The fact that this response was reversed during the summer suggests that the beneficial response of greater soil and root surface contact may be beneficial until environmental conditions become limiting for other essential growth factors under compacted soil conditions. As environmental conditions become more limiting, the reduced total root elongation rate and length in compacted soil ultimately decreases availability of water and nutrients by restricted access.

Shoot Responses

The direct and indirect effects of compaction on plant roots can elicit responses in the plant shoot as well. A number of studies on grasses demonstrate the potential of compacted soil to limit shoot density, verdure, clipping yield, soil cover, and lateral stem growth. Lower dry-matter production and yield under compacted soil conditions is attributed to reduced light interception caused by restricted leaf area development and is not a result of an impaired ability of crops to utilize intercepted radiant energy. Hormonal mechanisms have been proposed through which roots "sense" the mechanical impedance of the soil and elicit anatomical changes in leaf growth. An increase in abscisic acid in the xylem sap appears to be a root-to- shoot signal that lowers stomatal conductance and leaf expansion rates.

THE PRODUCTION OF CELLS

Moving fluids such as waterfalls, fountains, and the wakes of boats can generate specific forms. The study of the motion of fluid particles and the shape changes that the fluids undergo is called kinematics. The ideas and numerical methods used to study these fluid forms are useful for characterizing meristematic growth. In both cases, an unchanging form is produced even though it is composed of moving and changing elements.

An example of an unchanging form composed of changing and displaced elements in plants is the hypocotyl hook of a dicot, such as the common bean. As the bean seedling emerges from the seed coat, the apical end of the hypocotyl bends back on itself to form a hook.

The hook is thought to protect the seedling apex from damage during growth through the soil. During seedlin g growth (in soil or dim light) the hook migrates up the stem, from the hypocotyl into the epicotyl and then to the first and second internodes, but the form of the hook remains constant.

The hooked form is maintained over time, while different tissues first curve and then straighten as they are displaced from the seedling apex during growth. If a mark is placed at a fixed point on the surface, it will be displaced (indicated by the arrow), appearing to flow through the hook over time. If we mark a specific epidermal cell on the seedling stem located close to the seedling apex, we can watch it as it flows into the hook summit, then down into the straight region below the hook. The mark is not crawling over the plant surface, of course; plant cells are cemented together and do not experience much relative motion during development. The change in position of th e m ark relative to the hook implies that the hook is composed of a procession of tissue elements, each of which first curves and then straightens as it is displaced from the plant apex during growth. The steady form is produced by a parade of changing cells.

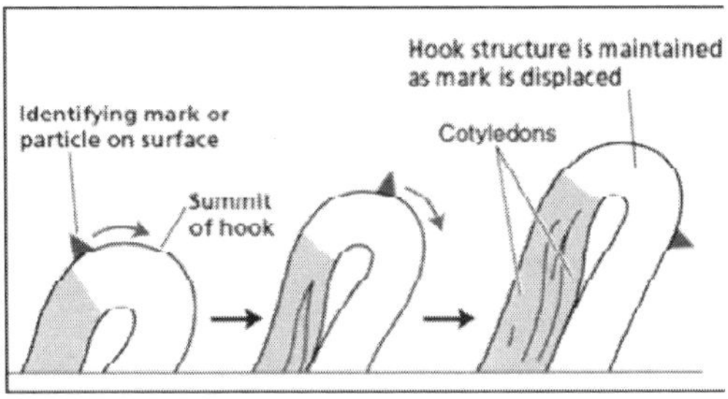

Fig. The dicot Hypocotyl hook is an example of a Constant form Composed of changing Elements.

A root tip is another example of a steady form composed of changing tissue elements. Here, too, the form is observed to be steady only when distance is measured from the root tip. A region of cell division occupies perhaps 2 mm of the root tip.

The elongation zone extends for about 10 mm behind the root tip. Phloem differentiation is first observed beginning at 3 mm from the tip, and functional xylem elements may be seen at about 12 mm from the tip. A marked cell near the tip will seem to flow first through the region of cell division, then through the elongation zone and into the region of xylem differentiation, and so on. This shifting implies that developing tissue elements first divide and elongate, and then differentiate.

In an analogous fashion, the shoot bears a succession of leaves of different developmental stages. During a period of 24 hours, a leaf may grow to th e sam e size, sh ape, an d bioc hem ic al com pos iti on tha t its neighbour had a day earlier. Thus, shoot form is also produced by a parade of changing elements that can be analyzed with kinematics. Such an analysis is not merely desc riptive; it pe rm its ca lcula tion s of the growth an d biosynthetic rates of individual tissue elements (cells) within a dynamic structure.

TISSUE ELEMENTS ARE DISPLACED DURING EXPANSION

As we have seen, growth in shoots and roots is localized in regions at the tips of these organs. Regions with expanding tissue are called growth zones. With time, meristems move away from the plant base by the growth of the cells in the growth zone.

If successive marks are placed on the stem or root, the distance between the marks will change, depending on where they are within the growth zone. In addition, all of these marks will move away from the tip of the root or shoot, but their rate of movement will differ depending on their distance from the tip.

From another perspective, if you were to stand at the tip of a root that had marks placed at intervals along the axis, you would see that all marks would move farther away from you with time. The reason is that discrete regions on the plant axis experience displacement as well as expansion during growth and development.

As a given region of the plant axis moves away from the apex, its growth velocity increases (the rate of elongation accelerates) until a constant limiting velocity is reached equal to the overall organ extension rate. The reason for this increase in growth velocity is that with time, progressively more tissue is located between the moving particle and the apex, and progressively more cells are expanding, so the particle is displaced more and more rapidly. In a rapidly growing maize root, a tissue element takes about 8 hours to move from 2 mm (the end of the meristematic zone) to 12 mm (the end of the elongation zone).

Beyond the growth zone, elements do not separate; neighbouring elements have the same velocity (expressed as the change in distance from the tip per unit of time), and the rate at which particles are displaced from the tip is the same as the rate at which the tip moves through the soil. The root tip of maize is pushed through the soil at 3 mm h-1. This is also the rate at which the nongrowing region recedes from the apex, and it is equal to the final slope of the growth trajectory.

THE GROWTH OF VELOCITY

The velocities of different tissue elements are plotted against their distance from the apex to give the spatial pattern of growth velocity, or growth velocity profile.

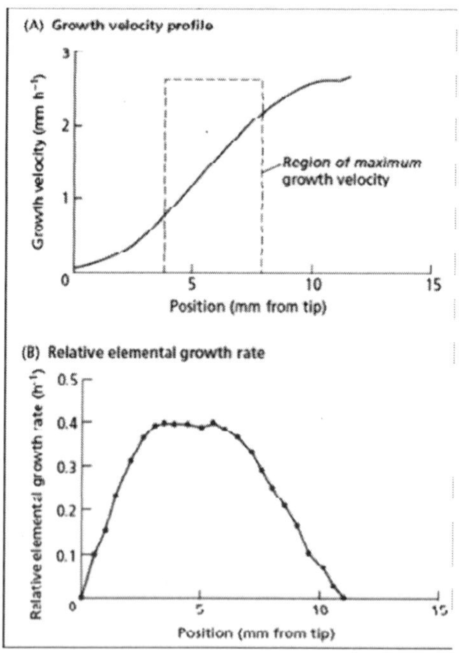

Fig. The Growth of the Primary root of Zea mays (maize) can be represented Kinematically by two related growth curves. (A) The growth velocity profile plots the velocity of movement away from the tip of points at different distances from the tip. This tells us that growth velocity increases with distance from the tip until it reaches a uniform velocity equal to the rate of elongation of the root. (B) The relative elemental growth rate tells us the rate of expansion of any particular point on the root. It is the most useful measure for the physiologist because it tells us where the most rapidly expanding regions are located.

Growth velocity increases with position in the growth zone. A constant value is obtained at the base of the growth zone. The final growth velocity is the final, constant slope of the growth trajectory equal to the elongation rate of the organ. In the rapidly growing maize root, the growth velocity is 1 mm h-1 at 4 mm, and it reaches its final value of nearly 3 mm h-1 at 12 mm.

If the growth velocity is known, the relative elemental growth rate, which represents the fractional change in length per unit time, can be calculated. The relative elemental growth rate shows the location and magnitude of the extension rate and can be used to quantify the effects of environmental variation on the growth pattern.

The Relative Elemental Growth Rate

One way to characterize the growth pattern along an axis mathematically is to plot the distance of a tissue element from the apex versus time. The resulting curve is called the growth trajectory. The slope of the curve at any given point is equivalent to the growth velocity at that region of the axis. With the apex as the reference point, the growth trajectory is concave. Marked tissue elements at different locations may be followed simultaneously to generate a family of growth trajectories.

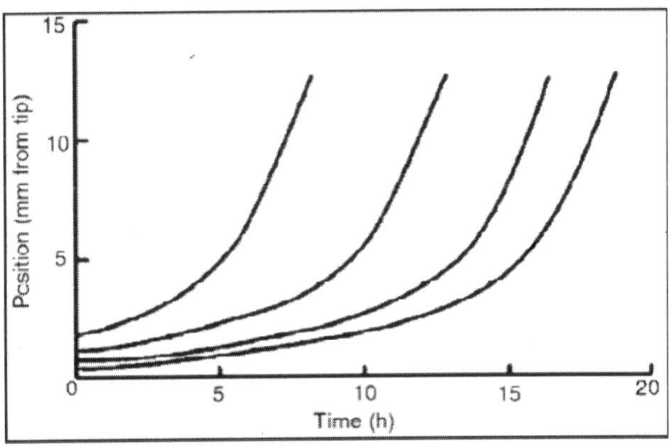

Fig. Growth Trajectories of Zea mays (maize): Growth Trajectories plot the positions of points spaced along the root axis as a Function of time.

Note that distance is measured from the root tip. This is termed a material specification of growth because real or material particles are followed.

As the regions of the plant axis move away from the apex, their growth velocity increases (i.e., the cells accelerate) until a constant limiting velocity is reached equal to the overall organ extension rate. The reason for this increase in growth velocity is that with time, progressively more tissue is located between the moving particle and the apex, and progressively more cells are expanding, so the particle is displaced more and more rapidly.

In a rapidly growing maize root, a tissue element takes about 8 hours to move from 2 mm (the end of the meristem) to 12 mm (the end of the elongation zone).

The root tissue element accelerates through this region. Beyond the growth zone, elements do not separate; neighbouring elements have the same velocity, and the rate at which particles are displaced from the tip is the same as the rate at which the tip is pushed away from the soil surface.

The root tip of maize is pushed through the soil at 3 mm h-1. This is also the rate at which the nongrowing region recedes from the apex, and it is equal to the final slope of the growth trajectory. The growth trajectory is a material description of growth because an individual element is followed through time. The growth trajectory provides a way to convert spatial information into a developmental time course.

For example, if the amount of suberin in the root increases with position, the growth trajectory will give information on the time required for the observed accumulation of suberin in cells moving between any two spatial points. To characterize the local expansion rate, we must consider the growth velocities at the apical and basal ends of a small segment of tissue. If the apical end of the segment is moving faster than the basal end, the segment is elongating.

The velocity difference, divided by the segment length, gives the relative growth rate of the segment. Now, imagine that the segment shrinks to a point x. To find the relative growth rate, r, at point x, we can use calculus: The value of r is given by the velocity gradient, dv/dx, where v is the growth velocity. This measure of the local growth rate is called the relative elemental growth rate. It represents the fractional change in length per unit time and has units of h-1.

If the growth velocity is known, the relative elemental growth rate can be evaluated by differentiation of the growth velocity with respect to position. Or we can approximate r by measuring the relative growth rates of closely spaced segments of initial length L. For each segment, r = (1/L)(dL/dt).

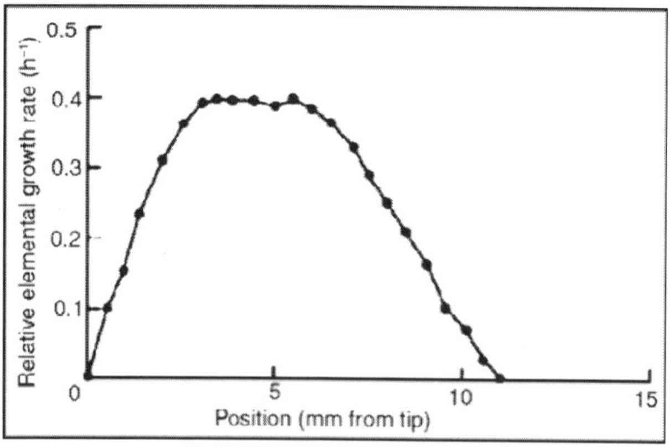

Fig. Relative Elemental Growth rate of Zea mays (maize).

The relative elemental growth rate tells us the rate of expansion of any particular point on the root. It is the most useful measure of root growth rate for the physiologist because it tells where the most rapidly expanding regions are located. The relative elemental growth rate profile, r(x), is a spatial description of the expansion pattern in the growth zone. In the maize root, the relative elemental growth rate has a maximum of 0.40 h-1 at 4 mm and drops to zero by 12 mm. That is, the tissue element at 4 mm is expanding 40% h-1, while tissue beyond 12 mm has ceased expansion. Again, to understand the profile in physiological terms, we must consider that individual cells expand more and more rapidly as they are displaced to 4 mm and then less and less rapidly as they are displaced through the basal part of the growth zone.

The relative elemental growth rate profile shows the location and magnitude of the extension rate and can be used to quantify the effects of environmental variation on the growth pattern. It is also the basis for studying the physiology of growth. The graph of r(x) can be compared to the distribution pattern of any substance thought to be regulating the growth rate. A close correlation between the two curves would provide circumstantial evidence for the role of the substance as a limiting factor for growth.

Embryo Maturation Requires Specific Gene Expression

The Arabidopsis embryo enters dormancy after it has generated about 20,000 cells. Dormancy is brought about by the loss of water and a general shutting down of gene transcription and protein synthesis—not only in the embryo, but also throughout the seed. To adapt the cell to the special conditions of dormancy, specific gene expression is required.

For example, the Abscisic Acid Insensitive3 (ABI3) and FUSCA3 genes are necessary for the initiation of dormancy and are sensitive to the hormone abscisic acid, which is the signalling molecule that initiates seed and embryo dormancy.

ABI3 also controls the expression of genes encoding the storage proteins that are depos ited in the cotyle dons during the ma turation phase of embryogenesis. The LEAFY COTYLEDON1 (LEC1) gene also is active in late embryogenesis. Because lec1 mutants cannot survive desiccation and do not enter dormancy, the embryos die unless they are rescued through isolation before desiccation occurs. The rescued embryos will germinate in culture and produce fertile plants, which are like wild-type plants except that they lack the 7S storage protein and they have leaflike cotyledons with trichomes on their upper surface. The normal appearance and development of the mature lec1 m ut a nts in dic a tes tha t th e LEC1 g e ne is r e qu ired o n ly dur i n g embryogenesis. Although the most obvious defects of the lec1 mutants are seen only in the maturation phase embryo, mRNA from LEC1 gene expression can be detected throughout embryogenesis. It has been proposed that LEC1 is a general repressor of vegetative development and its expression is necessary throughout embryogenesis.

CELL EXPANSION OF PLANTS

Within 5-10 minutes of application of auxins to isolated epicotyls, or intact tip of the plant or excised segments, growth in their length starts. This early effect is rapid and spontaneous. This rapid growth continues up to 30-45 minutes and then slows down but remains steady. This is also accompanied by an increase in cellular respiration, protein synthesis and the rapid extrusion of protons or in particular Hydrogen ions through the cell walls.

Another significant observation of this early rapid growth is insensitive to the inhibition of transcription (RNA synthesis) and translation (protein synthesis) - ex. Actinomycin-D at high concentration inhibits RNA synthesis completely and cycloheximide (CHI) inhibits cytosol protein synthesis. This clearly suggests that the early rapid growth requires neither the synthesis of RNA nor the synthesis of proteins. But this early effect of auxin is totally inhibited by colchicines at 0.5 to 1% concentration. Colchicine, an alkaloid from the tubers of Colchicum autumnale brings about the inhibition of tubulin (TB) polymerization into long microtubules, which act as the cytoskeleton for mechanical support in the cytoplasm. The colchicines effect on early growth further indicates the important role played by the assembly of microtubules in providing a mechanical structure and the force for all expansion.

Increased metabolic activity, particularly the degradation of osmotically inactive starch, etc., leads to the increase of osmotic pressure (they are related to each other) and a steep DPD gradient is created.

This facilitates the inward diffusion of water; consequently turgour pressure builds up within the cells. This force acts on all sides of the cell wall and forces the cell to expand in length and breadth. However recent investigations have shown that the cell elongation takes place under negative pressure.

The above features suggest that the process or the mechanism of cell elongation is complex and multiphased, but highly regulated. Various theories have been proposed from time to time and none of them could prove their claim because of one or the other drawbacks. Here we will restrict to a comprehensive account of cell elongation.

Early phase: The auxin's effect in the early phase on cell growth is rapid. This is insensitive to the inhibitors of transcription and translation but insensitive to colchicines and respiratory inhibitors like cyanide, DNP, etc., From this it can be deduced that cell elongation does not require new RNA and protein synthesis, but it requires ATP and ATP dependent tubulin polymerization into microtubular cytoskeleton. Auxin at the early stages activates and increases the rate of respiratory metabolism.

With the result ATP pool builds up rapidly. Auxin also activities membrane bound nucleating centre of tubulin polymerization.

These nucleating centres, utilizing ATP and tubulin monomers found in the cytoplasm, start polymerizing and soon build up long microtubules.

These, in turn, start orienting parallel to the long axis of the cell, which probably requires cytochalasin-B sensitive microfilaments.

With the rapid increase in microtubules, oriented in longitudinal plane, considerable force is exerted on the cell boundaries.

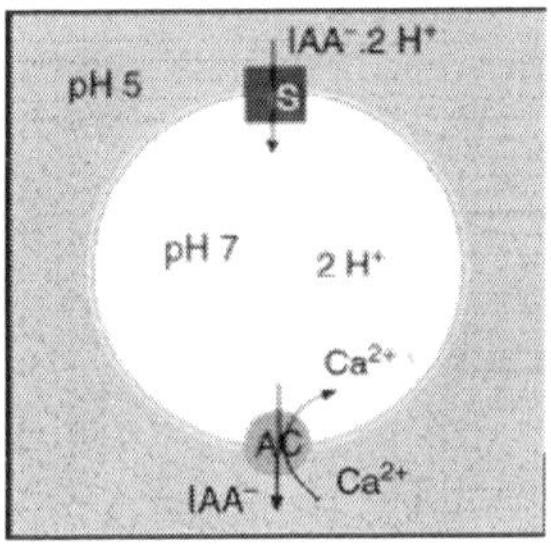

As the microtubular polymerization is building up, ATP ase/H* gets activated, and H* ions are pumped out into the interspaces of cellulose fibrils acidic pH in turn activates cellulases enzymes, which starts loosening up the cell-wall cellulose fibres. This loosening up of cell wall concomitant with acting as a mechanical force-the cell expands in length. This process starts 4-5 minutes after application of auxins and the elongation continues rapidly for about 20-30 minutes.

The late phase starts around 30-40 minutes after auxin application. This is slower than the first phase, but growth is steady. This phase is sensitive to Actinomycin-D, CHI, Colchicine and respiratory inhibitors, indicting that this process requires metabolic energy, tubulin-polymerization, RNA synthesis as well as protein synthesis. As tubulin pool gets exhausted in the first phase, it has to be replaced. That means it requires the synthesis of RNA and proteins.

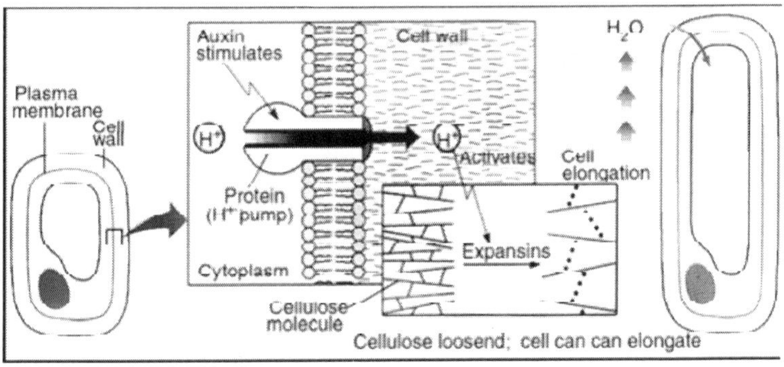

Apart from these proteins, the cytoplasm requires the synthesis of other proteins, for the increase in the mass of protoplasm. It is quite possible that the early activation of early translation may have significant effect on gene activation. Thus it takes some time for the second phase of growth to set in. The second phase provides all materials for sustained growth.

Growth Movements: The phenomenon of bending movement of plants in response to light stimulus is called Phototropism, the growth of roots towards soil is called Geotropism.

The curvature growth due to the touch is named as Thigmotropism, etc., Such differential growth movements are due to differential concentration of auxin. The stem tips and root tips have different optima of their auxin requirement to show maximum growth.

The mechanism of these movements will be explained in the chapter on growth movements. Auxin brings about phototropic movement because differential accumulation of auxin at exposed and darker side.

Root initiation: Vegetative propagation of stem cuttings is a normal practice in horticulture, floriculture and agriculture. Out of some stem cuttings which are put into soil, a few produce roots easily but others may or not produce roots. The roots that develop from any part of the plant body other than the radicle are called adventitious roots.

The development of such roots from the stem or any other part is a phenomenon of dedifferentiation. Stem is quite different fro the root with respect to its structure and function.

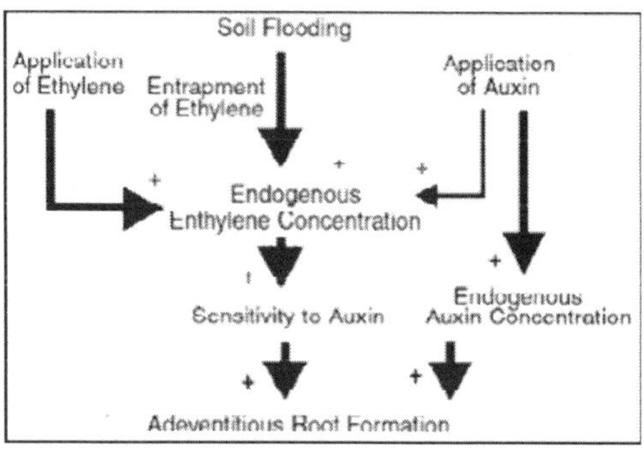

Still some of the tissues found in stem undergo dedifferentiation and develop into roots. This dedifferentiation and growth of roots from the stem is initiated by auxins Indole acetic acid, IBA, NAA, and other hormones. When applied to stem cuttings they induce the formation of roots. IBA has been found to be very good in this regard.This indicates that the pre-existing m RNA protein complex is drawn towards protein synthesis: it shows increased rate of protein synthesis. Some of the proteins thus synthesized in turn activate differential nuclear gene activation. As a result of it, a cascade of events occurs. Generally there is an increase in the synthesis of all cellular proteins. At the same time there is an enhanced rate of synthesis of some specific proteins. Also some new proteins are synthesized. Among these, tubulin, a membrane bound fraction, is synthesized at a high rate. It is believed that this protein along with microfilaments gets assembled into microtubules which orient themselves in a specific way so as to facilitate the polarity fixation for cell division.

In response to auxins, some of the cells in the pericycle region get stimulated and the required macromolecules are synthesized by differential gene activation. Thus these cells endowed with these new potentialities divide and redivide, and soon they organize into root primordial. They in turn develop into new roots which penetrate through cortical cells. The time taken for this new root formation, in response to auxin supply is just 24-36 hours. This effect of auxin induced new root formation has been well exploited by horticulturists and others. The following diagram shows how auxins interact with cytokinins in eliciting growth responses.

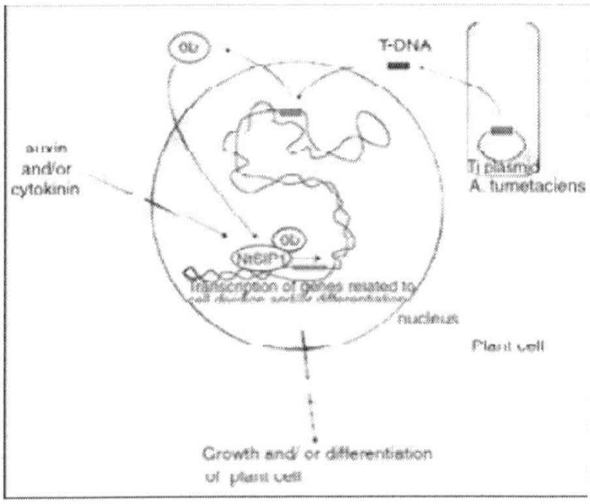

Apical dominance: Th e a ctive grow th of terminal bud inh ibits the development of lower axillary bunds. This effect is called apical dominance. Ma ny pla nts like conifers and other exhibit the apical dominance so predominantly that the plant grows tall with few branches and appear conical in shape. But plants like banyan, mango, tamarind do not show that much of apical dominance, hence, these plants spread outwardly. This phenomenon of apical dominance has been attributed to the differential concentration of auxin and nutrient supply to the axillary buds.

Abscission: With the onset of winter season the leaves of deciduous plants turn yellow and fall off. The falling of the leaves is due to the formation of abscission layers at the base of the petioles. This layer consists of thin walled but suberized cells. The middle wall gets disintegrated and the leaves fall down. The development of this layer is stimulated by a plant hormone called ABA. But the application of auxin to such leaves prevent the formation of abscission layer and thus prevent the falling off the leaves. This has a greater application particularly in preventing the premature falling of fruits like apple, orange, etc.

Parthenocarpy: Development of fruits without fertilization is referred to as parthenocarpy and such fruits are seedless. Normally plants do not produce parthenocarpic fruits. Normally plants do not produce parthenocarpic fruits.

But the application of auxin to flowers induces parthenocarpic fruit setting and prevents premature falling of fruits as well. Plants with triploid and many polyploid genomes do not set fruits. If such plants are sprayed with auxins at the time of flowering, fruit setting is enhanced. The fruits thus develope d are large, see dle ss, and sw eeter. Se edless ora nge s, lime, watermelon, sapota, ramphal and seethaphal, if they are produced this way, they have good market at the commercial level.

Gibberllic Acid

Gibberllic Acid is another natural hormone found distributed in higher plants. This was first discovered in Japan, later its structural and functional details have been elucidated by the people in the west. So far 52 or more different kinds have been identified. Among them GA2 2GA1 are the most common ones found in all plants.

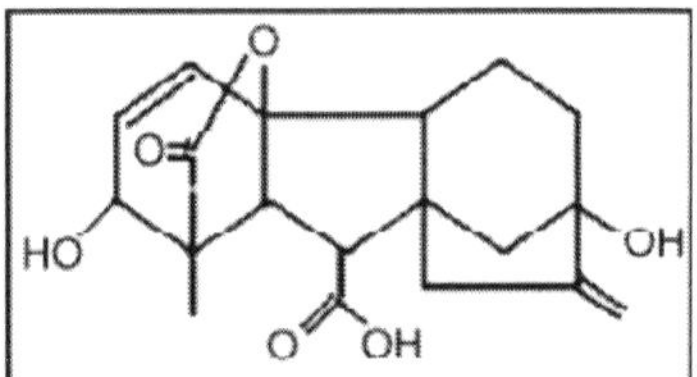

Gibberellic acid (GA_3)

GA is a steroid like molecule in its structure. Young leaves are the sites in which this hormone is mediated by phytochromes and then it is transported to the regions of growth.

The most significant effect of GA on the growth of plants is it effects internodal cell elongation profoundly than any other hormones. Even genetic dwarfs (eg. Dwarf Pisum sativum) grow as tall as the natural tall plants.

It also brings about a dramatic effect called Bolting and Flowering. Long day plants (which require 12-14 hrs of day light treatment for flowering) under short day conditions produce a cluster of leaves on highly shortened stem and never produce flowers. When such plants are sprayed with GA, th e s t em sta r ts elong a tion w ith e n h a n ce d gr owth of in tern od e s . Simultaneously the terminal buds starts producing floral branches. This phenomenon is called bolting and flowering. However both bolting and flowering are separated by time and space.

Besides these above said effects, GA is essential for the mobilization of food reserves for the germinating cereal grains like wheat and maize. At the time of germination, the young embryonic axis synthesizes this hormone, which is soon released into endosperm and reaches the aleurone layer. It is this layer, GA activates differential gene expression. As a result, specific m.RNAs is transcribed and later specific proteins like amylase are synthesized on translation. The amylase thus synthesized, acts upon the starch found in the endosperm, and the glucose thus released is drawn into the embryonic tissue which is in dire need of energy source. Notwithstanding the above effects, GA also effects parthenocarpy, rooting, germination, etc. This hormone also interacts with other natural hormones like auxin and cytokinin and brings about differential growth.

CYTOKININS

Another class of growth regulating hormones has been discovered by Miller et al and they are called Cytokinins. They are the derivatives or nucleotides. 6-Furfuryl amino purines, Benzyl amino purines and such compounds act as cytokinins.

Cytokinins are generally synthesized in the root tips, and then they are translocated to other regions. The endosperm of coconut milk, Borassus and of other plants is a rich source of cytokinins. Cytokinin level in plants is further regulated by other hormones like Auxins and GAs. On the contrary cytokinins control the synthesis of other hormones like Abscisic acid, a growth inhibitor.

Cytokinin per se has an important role in the cytokinesis. But along with auxins, it effects differentiation of plant organs from the callus. Cytokinin by itself inhibits root formation, but in the presence of auxins, it induces organ formation like roots and shoots. If the ratio between auxin and cytokinin is high, roots are differentiated, but the lower ratio induces the formation of shoots.

The intermediate ratio stimulates the development of these hormones in differentiation at the molecular level is very interesting by unfortunately it is least understood. Cytokinins are also found to break dormancy of winter buds as well seeds. The exact mechanism by means of which they over come the dormancy is not known, still it has been thought, that they effect through the suppression of IBA which is responsible for dormancy. Another interesting effect that the cytokinins produce is the prolongation of senescence (Ageing). This effect is often referred to as Richmond and Lang's effect.

❒

7

Growth, Development and Reproduction in Plant

The plant kingdom is divided into many classes and divisions. In 1883, Eichler separated the plant kingdom (Plantae) into two subkingdoms, the first being Cryptogamae (bacteria, algae, fungi, molds, liverworts, moss and ferns); the second being Phanerogamae (including gymnosperms and angiosperms). There have been many changes to the classification system, as we learn more about the genetics of different species. More recent taxonomy created new kingdoms that separated fungi, algae, and molds from what we commonly consider as plants. Plantae was separated into two divisions: Bryophytes (non-vascular) and Tracheophytes (vascular). Of the vascular plants, angiosperms (the flowering plants) make up approximately 300,000 of the 312,000 known number of living species. The number of 312,000 is probably a gross under-estimation. For anyone interested there is an estimated 20,000 species of non-vascular plants.

There are many different methods of reproduction that plants have developed over the centuries. If a plant's method of reproduction makes it easier to cross species or hybridized, it is more likely that its offspring will be better adapted to a changing environment For more than 8,000 years, human beings have also been affecting plant development by purposefully breeding new characteristics into plants, creating new species. The most important fact regarding reproduction is that every living organism has, in its cell's nucleus, a set of chromosomes. Chromosomes carry genes that have all the information necessary to make new individual organisms.

DIAGRAM OF A DICOT PLANT

One of the simplest methods of reproduction is when one cell divides and becomes two equal halves that will grow large enough to split again. With plants and animals that reproduce this way, each generation is identical to the one before.

CELL DIVIDING DIAGRAM WITH ONE PARENT CELL

Plant reproduction may be sexual, in which two parents produce a genetically different individual; or asexual, involving the propagation of plants that are genetically identical to the parent.

The Sexual Plant Life Cycle

Sexual reproduction is important in providing genetic variability. All plants that reproduce sexually must go through meiosis. Meiosis is a unique kind of cell division during which the paired sets of chromosomes present in sexually mature plants, called sporophytes, are halved. Because they have the two sets of chromosomes, one from the male parent and the other from the female, the cells of sporophytes are called diploid.

During this process, pairs of homologous, or identical, chromosomes, one from each parent, line up together. Crossing-over, or the exchange of genetic material between these homologous chromosomes, may occur at this time. Crossin g-over is critical for producing some of the gen etic variability in resulting offspring. Meiosis typically produces four haploid cells, each with one set of chromosomes, from a single diploid cell. These haploid cells become gametes-eggs and sperm. Without meiosis, sexually reproducing organisms would not have a mechanism for reducing the total ch romosome n um ber by ha lf so tha t genetic va ria bility c an be introduced via crossing-over. Meiosis is a critical step that must occur pr i o r to t h e f u s i o n o f s p erm an d egg , whi c h r e st o r e s the dipl o i d chromosome number.

Alternating Generations

Plants undergo a two-phase cycle of sexual reproduction known as the alternation of generation. This sexual life cycle involves alternation of the diploid sporophyte generation with the haploid gametophyte generation. The sporophyte stage begins with fusion of an egg and sperm, which produces a zygote. The diploid zygote develops into a sporophytic plant. This sexually mature plant eventually produces meiocytes, cells that will undergo meiosis, typically resulting in four haploid cells. These haploid cells, in turn, develop into gametophytes, which eventually form sex organs where eggs and sperm are produced. The sporophyte and gametophyte stages usually are easy to tell apart; one is typically parasitic or dependent upon the other. Which generation is dominant differs in different plants. Generally, the gametophytes of primitive land plants such as mosses are dominant and larger than the sporophytes. In higher, or more advanced plants, the reverse is true. In other words, the alternation of generations is the alternation between meiocytes undergoing meiosis and the fusion of gametes that produces a zygote.

The Gametophyte Stage

The gametophyte generation normally begins with a spore (sexual spore or meiospore) and ends with a gamete. The cells of the gametophyte are usually haploid.

The Sporophyte Stage

The sporophyte generation usually begins with a zygote and ends with a meiocyte. The cells of the sporophyte are usually diploid.

Reproduction in Fern

Sporophyte development is very rapid in common garden and woodland ferns. The zygote develops an embryonic root, leaf, stem, and foot. The foot apparently aids in absorbing nutrients from the gametophyte. The stem develops slowly, but ultimately produces leaves. The first few leaves of juvenile ferns differ from those of the mature sporophyte. As growth continues, the leaves finally begin to resemble those of the adult species.

Life Cycle of a Fern Diagram.

When the minute sporophyte has become established, the gametophyte dies. The gametophyte generation in ferns, as in other seedless vascular plants, is relatively short-lived and simple. The sporophytic phase is dominant in the life cycle. As in all vascular plants, what we recognize as "the plant" is the sporophytic generation, and at maturity it develops sporangia, the spore- producing organs.

In the common ferns, sporangia are borne on the undersides of fronds, or on modified fronds. The first fronds produced in spring are sterile-that is, they lack reproductive structures. The fertile fronds arise later in the season. In some species, such as lady fern (Athyrium filix-femina) the sterile and fertile fronds are similar. In other species-sensitive fern (Onoclea sensibilis), for example-the sterile and fertile fronds are very different. The fertile regions on fronds that bear sporangia are known as receptacles, and the group of sporangia on a single receptacle is called a sorus. In many species, but by no means in all, the sorus is covered during development by a flap of tissue called an indusium. The shape and arrangement of the sori are different in different species.

Spores arise by meiosis from sporocytes. For the most part, immature sori are a pale, whitish colour, although there are exceptions, such as autumn fern (Dryopteris erythrosora), whose sori, which are produced in the fall, are bright red. As the spores inside each sporangium mature, they typically get darker, until they are a deep brown or black. Not all spores mature to a deep brown or black, however. Polypodiums are ripe when they are buttercup yellow; osmundas, when they are green. The sporangia of ferns is typically a thin-walled case, usually on a stalk, and surrounded by a ring of thick- walled cells known as the annulus, which aids in opening the sporangium when the spores are mature.

The indusium, if present, ultimately shrivels when the spores have matured, exposing the sporangia. The annulus breaks near the base of one side, tearing the sporangium apart, and arches backward. Then the annulus snaps forward abruptly, catapulting the spores into the air. Ferns produce tremendous numbers of spores, but the special requirements of most species for moisture and shade effectively reduce the number of gametophytes that ultimately develop. Spores that are deposited by air currents on sufficiently moist soil and rocks germinate within five to six days, and the gametophyte begins to develop. Most fern spores also require light for germination; those of the bra cke n f ern (Pter id ium a quili num) ar e a n ex ceptio n. Earl y in germination, colourless rhizoidal cells form at the base of the filament of green cells. These rhizoids, or root-like hairs, absorb and carry water and nutrients to the developing gametophyte. Eventually, the gametophyte becomes heart-shaped, and in some species may be as big as half an inch in diameter. The fern gametophyte was long ago called the prothallus or prothallium, because it was known to be the precursor of the fern plant, even before its sexual function was clearly understood.

As well-n ourish ed, bise xual ga metoph ytes ma ture th ey develop antheridia and archegonia, normally on their undersides. The female, egg- producing organs, or archegonia, occur near the notch of the heart-shaped prothallus. The male, sperm producing organs, or antheridia, occur on the "wings" and opposite the notch. Each archegonium has a chimney-like protuberance that is flared at the top to receive the sperm. Each antheridium is a minute, capsule-like sac where the sperm grow. The cells inside the neck of a mature archegonium dissolve, creating a moist passageway through which the sperm can swim to the mature eggs. The eggs of several archegonia may be fertilized, but usually only one of the zygotes develops into a juvenile sporophyte.

It is interesting to note that while bisexual prothalli may be capable of fertilizing themselves, experiments have shown that the resulting sporophytes often fail to thrive. This suggests that cross-fertilization, the union of sperm and egg from different prothalli, necessary for increasing genetic variability within species, is promoted in at least some ferns.

Asexual Reproduction

Asexual propagation can occur in various ways. Many plants produce genetically identical copies of themselves, through a mechanism referred to as asexual reproduction. However, botanists more properly refer to this mechanism as "sexual propagation" or "vegetative propagation" because many hold that only sexual reproduction should be referred to as true reproduction, since, this is the only kind of propagation that results in the production of new, genetically unique individuals.

Another type of asexual propagation occurs when plants develop underground stems (or rhizomes) which grow outward, or new shoots which grow upward to form new shoots that are genetically identical to the parent. One such example is the trembling aspen (Populus tremuloides), which sometimes deve lops en tire stands of trees growin g out of the ground as seemingly individual stems, but are actually genetically identical and interconnected below-ground. Another example is the strawberry (Fragaria virginiana), although this species has its vegetative runners above- ground.

Other plants develop bulbils on their stems, which can detach, fall to the ground, and sprout to develop new plants that are genetically identical to the original one. One familiar species that does this is the tiger lily (Lilium tigrinum. Other plants can propagate from twigs or stem pieces that fall from the parent, then lodge into a suitable site and develop into a ne w plant. The c rack willow (Salix fra gilis) ca n spre ad itself a long watercourses in this manner (as well as by disseminating seeds). Other non-sexual means of propagation include the production of underground bulbs, corms, and tubers that split into parts, each of which is capable of developing into a new plant. Plants that reproduce in this manner include irises and daffodils. All plant organs have been used for a sexual reproduction, but stems are the most common.

Stems

In some species, stems arch over and take root at their tips, forming new plants. The horizontal above-ground stems (called stolons) of the strawberry (shown here) produce new daughter plants at alternate nodes.

Underground Stems

- Rhizomes
- Bulbs
- Corms and
- Tubers

Irises and day lilies, for example, spread rapidly by the growth of their rhizomes.

Leaves

This photo shows the leaves of the common ornamental plant Bryophyllum (also called Kalanchoë). Mitosis at meristems along the leaf margins produce tiny plantlets that fall off and can take up an independent existence.

Roots

Some plants use their roots for asexual reproduction. The dandelion is a common example. Trees, such as the poplar or aspen, send up new stems from their roots. In time, an entire grove of trees may form—all part of a clone of the original tree.

Plant Propagation

Commercially-important plants are often deliberately propagated by asexual means in order to keep particularly desirable traits (e.g., flower colour, flavour, resistance to disease). Cuttings may be taken from the parent and rooted. Grafting is widely used to propagate a desired variety of shrub or tree.

All apple varieties, for example, are propagated this way. Apple seeds are planted only for the root and stem system that grows from them. After a year's growth, most of the stem is removed and a twig (scion) taken from a mature plant of the desired variety is inserted in a notch in the cut stump (the stock). So long the cambiums of scion and stock are united and precautions are taken to prevent infection and drying out, the scion will grow. It will get all its water and minerals from the root system of the stock. However, the fruit that it will eventually produce with be identical (assuming that it is raised under similar environmental conditions) to the fruit of the tree from which the scion was taken.

Apomixis

Citrus trees and many other species of angiosperms use their seeds as a method of asexual reproduction; a process called apomixis.

- In one form, the egg is formed with 2n chromosomes and develops without ever being fertilized.
- In another version, the cells of the ovule (2n) develop into an embryo instead of—or in addition to—the fertilized egg.

Hybridization between different species often yields infertile offspring. But in plants, this does not necessarily doom the offspring. Many such hybrids use apomixis to propagate themselves. The many races of Kentucky bluegrass growing in lawns across North America and the many races of blackberries are two examples of sterile hybrids that propagate successfully by apomixis.

- The female cones of their own species (rare) or
- Those of a much more common species of cypress.

Breeding apomictic crop plants

Many valuable crop plants (e.g., corn) cannot be propagated by asexual methods like grafting. Agricultural scientists would dearly love to convert thes e plants to apomixis: makin g embryos that are genetic clones of themselves rather than the product of sexual reproduction with its inevitable gene reshuffling. After 20 years of work, an apomictic corn (maize) has been produced, but it does not yet produce enough viable kernels to be useful commercially.

Factors affecting plant reproduction

Climate, soil moisture, soil nutrients, disturbances (such as defoliation, fire, cultivation, and disease) are some of the more common things that affect the reproductive ability of plants. Tolerance versus Requirements: not all plants respond the same to stress. For example, both Agrostis tenuis and Festuca ovina have the same tolerance limits for soil pH, but in a stand of Agrostis and Festuca, Agrostis had a maximum abundance at pH 4-6 and Festuca had a maximum abundance at pH 3-4 and 6.5-7.5. There is also a difference within species. For example, 2 ecotypes of Dactylis glomerata (Norwegian and Portugese) have different temperature requirements for reproduction. In Portugal, it is a winter-grown crop (lower temps) and in Norway it is a summer crop (higher temps) and they respond differently when grown in each other's typical environment.

Plant Community Development

Plant Community - "a group of plants living under relatively similar conditions in a definite area"

Development-"to expand or bring out the potentialities; to advance from a lower to higher state" If these two are combined, then plant community development could be considered "to expand or bring out the potentialities a group of plants living under relatively similar conditions in a definite area".

Stages of Community Development

There are numerous papers on the stages of community development: Succession. Nudation, migration, ecesis (adjustment of a plant to a new habitat), competition, reaction and stabilization.

Disturbance

When discussing community development, it implies that there has been some sort of disruption in the vegetation of the community. Grime defined disturbance as "the mechanisms, which limit the plant biomass by causing its partial or total destruction". Disturbance should always be specified according to spatial extent, temporal extent and magnitude.

Type of Disturbance

Grubb summarized information for different types of plants under different disturbances. While he thought it possible to generalize about type of disturbance and plant life form it was difficult to generalize life characteristics and high disturbance levels. For example, demographic characteristics of communities exposed to grazing reveal a high number of perennial plants capable of both sexual and vegetative reproduction. Annuals and biennials are generally located in interstices and fluctuate in relative abundance.

Grazing intensity tends to increase the dominance of clonal plants. The number of species with vegetative reproduction were significantly different under grazing than those with vegetative reproduction. In the same study, seed dispersal was also analysed. Mobile and adhesion, as methods of dispersal, were both significantly different under grazing. Wind dispersal was significantly different under soil disturbance.

On a denuded glacial foreland, pioneer species were plants whose dominant reproductive form was vegetative. Unlike early colonizing species of secondary successions, pioneers of primary succession are usually long- lived perennials that are relatively slow growing.

Research by Vose an d White in a burn ed Ponderos a pin e (Pinus ponderosa) commun ity indicated that buried seed did not contribute significantly to seedling recruitment, and seed rain was more important than buried seed. Most regeneration was from seed rain, and the only vegetative reproduction was from the shrub Ceanothus fendleri. It should be noted that this shrub also reproduced by seed. Allogenic disturbances reveal a question to consider: can we take research from autogenic disturbances and extrapolate them to allogenic disturbances?

Intensity and Frequency of Disturbance

Key characteristics suited to intensity of disturbance are very different. Grubb thought special attention should be placed on plants that were able to increase their range after a disturbance and more study was required on life form, phenolgy, and requirements and tolerances.

Vose and White found that intense heat from a pole-stand burn reduced seed rain compared to sites that burned with less intensity. Burning prevented germination of any buried seed. In their study, the authors speculated that post-fire environmental conditions might ha ve been unfavourable to germination and establishment. It is commonly thought that ruderals, with high seed production, generally dominate areas of frequent, high disturbance. Seeds are usually dormant, and these plants are largely opportunistic of the increases in light supply and nutrients in a disturbed site. A summary of data by Grubb in 1985 supported this theory. In natural areas of yearly disturbance, (for example, scour of drift lines by floods), there is a high number of annuals that use the abundant nutrient supply and set seed in the first season. The seeds of these annuals have no specialized means of dispersal and rely on dispersal though the drift material.

Environmental Conditions: where did the Disturbance occur?

Low Temperatures

On the tundra, there are very low reproductive rates and predominance of vegetative proliferation. Most invading species are either self-pollinated, or invest massive resources in vegetative propagules there is a higher representation of species that possess some mode of uniparental reproduction.

Chapin calculated that the carbon cost of producing a tiller of Dupontia fisheri from seed at Pt. Barrow, Alaska was 10,000 times greater that that of producing tiller vegetatively. A single Betula pubescens in Swedish Lapland produces 40,000,000 seeds thoughout its life, only one of which is necessary to replace the original tree for a stable population. Are these seeds only important in the colonization of new areas outside the forest or after disturbance such as fire within the forest? Recruitment from seed is important in open habitats such as fell-fields, but early mortality rates are high. Vegetative reproduction has low recruitment levels but high survival rates due to the physiological interdependence of modules. (fell fields are open communities of scattered individuals aggregated into islands). The plants are dwarfed and prostrate, and the habitat is characterized by little winter snow cover, water shortage, primitive soils and frequent soil movement due to freeze-thaw cycles).Chambers cautioned in diminishing the importance of sexual reproduction in tundra environments. And though compared to temperate climes, sexual role is reduced, for tundra it is still very important.

Callaghan and Emanuelsson indicate that sexual reproduction may be adapted to the exceptional and not too normal circumstances. Environmental conditions that select for sexual reproduction may be very different from those that select for vegetative growth.

Low nutrients

On the tundra and in alpine disturbances, distinct species colonization patterns can be observed. Chambers noted that species with nitrogen-fixing symbionts are frequent colonizers on mineral soils low in nitrogen, including those exposed by retreating glaciers or snowbanks. In alpine tundra areas of severe soil disturbance, where the resultant soils are poorly developed and have low nutrient status, colonization is most frequently from dissemules, mostly seeds dispersed from adjacent communities. In these conditions, it is necessary for plants to produce many seeds that can easily be dispersed

Importance of Soil and Water

Prach and Pysek studied the role of clonal plants in plots that were 10 to 50 years old, grown under different types of habitats. They had acidic emergent bottoms, peaty soils, sandy, reclaimed, abandoned sand pit, ruderal sites ranging from poor to rich and old fields that were wet to xeric. From their observations of old fie lds, clonal plants dominated under xeric conditions, while non-clonal dominated under mesic conditions. These sites were 50+ years old. This indicates that more long-term studies in old fields are needed under various environmental conditions.

The study they did for plots that were 10 years old concluded the data set was limited and not able to reveal a functional relationship between the participation of clonal plants and habitats.

Stocklin and Baumler found that seeds found in areas around a glacier (primary to secondary successional areas) were extremely dependent on water ava ilability for germ ination and establishment. Seedling success was dependent on "safe" sites, which were slight indentations or areas covered by dead plant material. Chambers found that in windy, soil exposed areas (tundra/alpine) relationships between soil characteristics and seed attributes significantly affected the spatial distribution of dispersed seeds. This may be one of the primary determinants of species colonization patterns.

Chambers found that with small soil particles, small seeds and seeds with adhesive coats were trapped, but most large seeds moved across the surface; at large particle size, larger seeds were trapped, with larger particles, small seeds moved further downward in the soil as well. In the same study, surface soil temperatures were measured and on dark covered organic soils, the temperature ranged from 35-45 °C on the surface and 20-25 °C at 5 cm depth. On an adjacent site with exposed, loamy sands, midda y soil temperatures were about 5 °C lower. Soil temperature can result in higher rates of mineralization and decomposition. Most tundra species had higher germination rates under higher temperatures than what is normally found in the field.

Competition

Small clones spread laterally and are not overtopped by larger clones. Dominance and suppression by non-clonal Genetic variability in plants

Temporal Considerations

Noble and Slatyer, listed the following three plant attributes as most important in vegetation replacement:

- The method of arrival or persistence of the species at the site and after disturbance.
- The ability to establish and grow to maturity in the developing community.
- The time for the species to reach critical life stages.

Epp and Aarssen found interactions between species in recently established vegetation (1 year) are affected more by differences in attributes that are relevant to early life, i.e seed weight. This relates to the competitive ability of seeds in establishing themselves in an open area. They also discovered that the older a stand became, other attributes became important in a plant's ability to remain dominant. In a 1 year old plot, lateral spread only accounted for 0.3% of the variation, while in the 11 year old stand, it accounted for 26% of the variation in species diversity. Seed weight (number of seeds was not measured) accounted for 13% of the variability in the 1- year-old plot, but only 0.03% in the 11 year old plot.

Non-clonal plants had maximum coverage after 4 years, and the clonal plants had maximum coverage in years 6 and 7. Coverage for the clonal plants decreased after year 7. They also determined a difference between guerilla and phalanx plants. Phalanx had a peak of 70% coverage in year 5 and that decreased to 28% coverage in year 10. Guerilla plants peaked at year 6, at 72% and in year 10 were 61% (this was found to be significantly different).

Species capable of guerilla growth are often more successful in the later stages of succession than plants with phalanx growth; this may be explained by the guerilla's ability to produce widely spaced modules that have a greater chance of penetrating into more or less closed vegetation. Differences in clonal growth forms were also found by Oborny in simulation models of different types of vegetatively reproducing plants. Clonal plants that had a rigid structure decreased in number over time vs spp that had plastic growth form.

Vegetative reproduction becomes of more importance in later stages, when dense cover and a thick litter layer limit establishment from seed. But there are limitations to vegetative reproduction as well. Belsy found that populations of Sporobolus fimbriatus and Pennisetum mexianum began to replace Digitaria maroblephara and Sporobolu ioclados after 5 years of non- grazing. This is due to the fact that S. fimbriatus and P. mexianum are both tall rhizomatous species, whereas Digitari maroblephara and Sporobolus ioclados are both stoloniferous. The build up of the litter layer suspended the stolons above the ground. This was also true for species that reproduced by seed. The seeds could not germinate with the heavy litter layer.

Factors that Influence Colonization: Specifically in Regards to Sexual and Vegeta Reproduction

It has been hypothesized that the role of clonal dispersal in succession is evident both in early stages, due to the rapid capture of space, and in late- successional stages, due to the rapid filling of gaps in more or less closed vegetation cover. In es sen tia lly closed cover, s eeds do not h ave the opportunity to germinate, nor do seedlings have the chance to establish. Competition for nutrients, water and light is just too great. This would imply that eventually all plant communities should be dominated by plants that rely heavily on vegetative reproduction. Why is it then, that there are many plant communities that appear to stabilize with a combination of plants that use both seed (sexual) and vegetative (asexual) reproduction? Research on the role of clonal species is lacking, and an understanding of their behaviour is based to a large extent on speculation. The factors that inhibit or promote one reproductive strategy over another are not well understood.

❒

8

Selection and Crossbreeding in Plant

Here is hardly any plant-breeding project which does not involve cross-breeding. In most cases it is the object to produce variability, in order to obtain the material for selection to work upon. The quickest way to obtain variability is cross-breeding.

We should distinguish between the work of the plant breeder and that of the scientific worker. It is true that the great majority of plant-breeding problems are tackled at the larger institutes by genetically trained specialists, and yet the distinction should be made. In scientific experiments only the most painstaking precautions are good enough. We must be absolutely certain about the paternity in our hybrid plants, and for this reason both the collection of the desired pollen and the elimination of the pollen from the mother-plant must be done with the minutest precautions.

If our object is a wholly practical one, we may often relax such precautions to a certain extent. For instance, in some plants castration—the elimination of stamens—is not only extremely difficult, but the manipulation brings a risk of seeing the flower wither and die. In such plants (soybeans, oats) castration can often be dispensed with, when we choose the parents in such a way that we can recognize the hybrids from inbred plants.

In every plant-family our first step must be to study the biology of the flower, in so far as this has not been done by others. We must know at what stage of growth of the flower- bud the anthers dehisce, so as to hit upon the safe time at which we must perform the operation of removing them. We must also know when the stigma is receptive, to find the moment at which to apply the pollen.

In most plants castration and the application of pollen can be done at the same time. This is the case in rice, in most of the cereals, in tobacco, Ipomoea, cocoa, peas, beans and a host of other plants. In some plants with large flowers the stamens are attached to the corolla, and it is possible to cut off the corolla with the stamens in one operation. This can some- times be done with the fingernail (apple, raspberry, straw- berry) or with a scalpel.

When a scalpel is used, it is sometimes advantageous to cover the blade almost up to the tip with adhesive tape, to avoid cutting down too deep. Where this is impossible, the stamens have to be extracted one by one. A set of fine pincers with differently curved tips should be in the possession of every worker. Very useful fine instruments can often be made to fit particular occasions. A very fine hook, made out of a bent pin, and mounted into a needle-holder can sometimes be used to great advantage, especially in castrating such minute flowers as in alfalfa or clovers.

For fine work a watch-maker's eyeglass, or even a binocular dissecting glass, may come in useful. On the whole, dispensing with castration in plants with minute flowers, and to use the method of straight cross-breeding, leaving the finding of hybrids till next season.

In some plants male and female flowers are separate. In such plants no castration is necessary; all we have to do is to protect the flowers from unwanted pollen. In cucurbits, melons, squashes, cucumbers, we found it quite easy to close unopened flower-buds, by means of small pieces of copper, or preferably lead-wire. Of course both male and female buds should be so protected. Lead-wire has the advantage of being very soft, and it can be used many times in one season.

Especially in wind-pollinated plants, care should be taken to protect the pollen used against contamination with unknown pollen. For this purpose the flowers from which pollen must be collected should be bagged. In some plants pollen is so abundant that we can use a separate anther for every flower to be pollinated, and we can hold this with fine forceps. Where the pollen is dry, it can be shaken out of the anthers into some small receptacle, for which a small gelatine capsule is just the thing. Such capsules can be closed, and the pollen contained in them can often be kept in very good condition for days, if the labelled capsules are kept cool in a well-closed bottle over calcium.

In maize both the male and female inflorescence should be bagged early. When the silks protrude they can be trimmed off short, and the pollen shaken over them from a tassel. If it is wished to obtain a cob full of kernels, it is of some advantage to insert a flowering tassel into the bag that covers the cob, and we can prolong its life and usefulness by standing this tassel in a small bottle of water hooked to the stalk.

In certain plants the operation of castration and fertilization can be performed on cut-off stalks that have been taken into the labouratory. In sugar-cane, in millet and sorghum, and even in wheat and barley, such panicles and ears, when stood in water, will ripen their seeds well, especially if only a few flowers are left on them for this purpose. Paper bags should be of ample size, and it is necessary to leave them on no longer than is absolutely essential for full protection during the few days in which the stigma is receptive. In certain cases (coffee, cocoa) a finely woven tartan may have advantages over paper; it is softer, and can be arranged around awkwardly placed flowers better than paper. Of course it is useful in certain cases to give the ripening fruit or seeds some protection against loss or damage (stone- fruit, potato, tobacco, (Enothera), and paper bags may be necessary at that stage.

For keeping small quantities of seeds, small paper or preferably cellophane envelopes can be used. Gelatin druggist capsules, however, are preferable. The labels should be en-closed with the seeds.

LABELLING, NUMBERING AND GADGETS

Every plant-breeder and every geneticist has his special methods of keeping his records. To be able to do this he must have suitable methods of numbering or labelling his plants or seeds.

Any system of labelling or marking must be legible and indestructible by weather or insects. For merely marking, as an aid to tracing a special plant or part in a field, a coloured thread may be quite sufficient. If such a thread is used in addition to a number or a label, it helps to distinguish that label. Paper labels, even if made from very good parchment paper, tend to become discoloured.

To depend on such labels only, let us say in marking pea-pods, is courting trouble, for they closely resemble dried leaves or empty pods. The well- known Vilmorin red woollen thread is of very great value; it may even serve the double purpose of protecting the cross-fertilized ear of wheat or barley and of making it easy to find it again.

To mark seed capsules, fruits and such things, we found that lead-wire was much better than string. We can string a glass or china bead on a small piece of very soft lead-wire, and if we need a few dozen distinguishing marks, combinations of coloured beads will be excellent. We used this method in marking cross-fertilized squash blossoms and melon blossoms. If we make a loop of sufficient width, we can squeeze one end of this against the fruit- stalk, and it will accompany the fruit until harvesting time, or even beyond.

A very good label that will never lose its legibility is a poultry leg-band with a perforated number. These are made out of thin aluminium, and can be opened and adjusted before closing. Such labels are just the thing to mark plants that must be taken up and wintered in a cellar, like dahlias. Most seedsmen use stout painted stakes for carrying the name and the number of rows and individual plants, the name facing the row it marks. To obviate mistakes it is safer to have duplicate stakes, or to have the name and number of each lot written on a separate stake.

To save time and worry, good use can be made of poultry leg-bands or of the newer coloured plastic numbered wing- shields. These are numbered on both faces, and can be bought in consecutively numbered hundreds or thousands.

They can be hung from wire stakes with eyelets, and they accompany such plants as cereals or beans when they are bundled for harvesting. In some countries one has to guard against theft and mischief. A European blackbird pull out dozens of numbered yellow wooden labels and scatter them all over the place. A system of book-keeping which allows one to find the missing numbers from the book is indispensable if we want to avoid mistakes.

Numerous methods have been invented for protecting flowers and branches against insects or wind-borne pollen. Paper bags come first; they must be strong, rainproof and light. Some cotton-wool may come in very useful in making the connection of bag and stalk pollen-proof. Glassine bags and lengths of seamless sausage casing are being widely used in the United States. They allow one to see the inside of the bag, and so may save much time and error.

In some sympetalic plants flower-buds may be closed hermetically by the application of a length of very soft wire (lead- wire) wrapped round the bud, and used again after fertilization. We have used this method with squashes, melons and Ipomoea, The American plant-breeding journals and publications abound with methods for castrating and cross-breeding plants. Removing the pollen from the stigmas by washing it off or killing it by heat are methods but they sound excellent.

Gelatin capsules, made in two pieces, of which one fits over the other, are handy receptacles for keeping very small quantities of seeds or of pollen, for slightly larger quantities sachets for internal use may be used; they are cheap, and can be written upon, but if we close them we can never use them again. Paper bags must be used a good deal in storing seeds and similar objects (bulbils, bulbs, corms). They can be closed in different ways. The simplest system is the method of the French seed firms, in which each bag is closed by three folds, in such a way that it will never open spontaneously, even if submitted to considerable shaking.

The gadgets used in castration and cross-pollination are very numerous, and must be adapted anew to every new plant. Of course it is necessary to kill adherent pollen with alcohol, and extra scalpels, forceps or scissors are of great help. In working with very minute flowers a good binocular microscope is invaluable.

Avoid castrating minute flowers as much as possible. Everyone doing this sort of work occasionally regrets he was born with only two hands and without a prehensile tail. A ring to which has been soldered a minute cup that holds a small gelatin capsule with pollen, and that can be worn on the index-finger of one hand, is almost as useful. This is another American invention.

Hard seeds can be treated in different ways to make them grow. Some seeds, like those of canna or musa, can be scratched with a file, others may be rubbed on stone. Enothera seeds can be soaked in water; some seeds will only take up water rapidly when we exhaust the air above the water in which we keep them and then readmit the air.

Seeds of apricots, cherries and plums can be extracted after carefully cracking the stone. The extracted embryos (and the same is true for embryos of apples, pears and citrus) must be protected against bacteria and mould. Thymol solutions, even weak ones, will do this without injuring the seedlings.

SPEEDING-UP PLANT-BREEDING WORK

Plant-breeding work involving crosses generally takes a great number of generations before the superior new plant has been obtained. In the routine work of a large plant-breeding station so many breeding projects arc under way simultaneously that it matters little how much time elapses between the initial cross and the final purification of a promising commercial variety. Every year there are many projects being started, several are under way, and a few arc in their last stages.

It often pays to cut down the time required for work of this kind. One of the most promising means of doing this is to grow more than one generation of plants in one year.

This usually means that one of those generations of plants is grown in a season in which the plant is not ordinarily grown, and that for this reason a correct evaluation of the quality of the individual plants is out of the question.

In Europe or in America growing an extra generation on the spot means growing plants in the greenhouse. To grow cereals or beans in winter in greenhouse conditions not only requires heat; we must also have the means of regulating the length of day. The simplest method is to use large electric lights to furnish both the heat and the light, to lengthen the day by those means, and to lẹt the temperature go down during the dark part of the night. Under those conditions cereals and beans will flower and set seed very well in the winter months,

The electricity required per square yard is roughly that used by lamps that require a kilowatt. It is clear that under such conditions the winter generation will not show the normal qualities of the plants in that group which are grown in summer-time in the field, but there are many cases in which this is not of much importance.

If we want to proceed by the method of making some hybrids and then growing a second and third inbred generation from this stock, we can do our crossing in the field, grow our hybrids in the winter, and then grow another crop of second-generation plants in normal field conditions. When we are using the method of mating back hybrid stock to some pure kind of plants we want to improve it matters but little if we cannot very well select the best plants during some of the generations involved.

It is evident that we must grow a generation under normal field conditions whenever we want to make our selections. In recent years a few seed-firms have hit upon the perfect scheme of growing two generations of annual plants. This does not involve any greenhouse generations. They work in two different localities with very similar climate, but so chosen that one of those localities is situated in the Southern Hemisphere.

There are regions of Chili where Californian firms can find conditions of plant-growth that are identical with conditions at home, with the exception that the growing season comes during the Californian winter. In those circumstances it is quite possible to grow two absolutely normal generations during one year. Our present greatly improved methods of communication make such schemes perfectly feasible.

With many tropical plants one generation yearly is grown of plants that need special conditions of either a wet or a dry season. Here it is easy enough to interpolate an additional generation, by looking for a suitable spot, or by providing the necessary irrigation.

Sometimes it takes a very long time to bring the plants to a condition where we can evaluate them. A good example is that of the fruit trees. It takes many years for an apple seedling or a seed-grown cherry to come into bearing. Here we can save a great deal of time by budding or top-grafting parts of the seedlings upon mature trees in full bearing. By this method we may save more than half the number of years normally required for bringing the seedlings to maturity.

In this connection we might also treat of those cases where certain seeds take a very long time to germinate. Storing the seed in suitable conditions of moisture, and especially of temperature, will speed-up germination in such plants as gooseberries and celery. In stone-fruits it has been found possible to crack open the stone, to extract the kernel and grow the seedlings under aseptic or even under antiseptic conditions (thymol solution).

SHOWS AND SHOWING

The exhibiting of plants and seed at the shows is a very curious thing. It pays extremely well for a seed-firm to employ some gardeners who know just exactly when and' how to get enough plants full of luscious beans or covered with flowers on the exact day for each show.

Florists and lovers of gardening see the quality of such plants at the shows, and many people who admire them there try in their turn to win the coveted prizes at their local shows the next year. This show game is certainly very nice and pleasant; it gives the gardeners something to strive after; it may even help to keep some of the lads in the country.

On the other hand, there is very, very little connection between the qualities judged at the shows and actual value of the material shown. The few selected mangolds, or the mammoth pumpkins, or the bundle of wheat- ears, or even of the maize-ear contests of the American agricultural shows.

All along the line, in horticulture as well as in agriculture, the selected samples shown give no indication of the value of the seed or plant stock the producers have for sale. Of course we can see that those few mangolds have the ideal shape and size, but the seed-firm must give us some guarantee about the percentage of mangolds that will grow to this size and shape from the seed they sell, just as the owner of a prize-winning bull must give some guarantee of this animal's ability to give profitable daughters, in addition to the blue ribbon awarded to the beautifully coloured and shaped prize-winner.

In agriculture this is beginning to be so well realised that the classes for beautiful corn cobs or for the largest mangolds are becoming very rare indeed, even if we still see stock-judging in dairy cows and similar farces. In the horticultural shows much interest is still shown in regard to flowers and vegetables.

It seems evident that when we strive for beauty, beauty contests for roses and begonias, for asparagus and delphiniums and dahlias are indicated.

The show is a shop-window. The seller must show if his competitors show, if only to keep his name before the public. To the public the show is only just a pleasant occasion for a half-day outing. To the buyer the show is much nicer to see than a description in a catalogue.

Long lists of novelties are bought at the shows by people with an experimental turn of mind. We see the representatives at the shows kept busy noting down long lists of orders. Showing certainly pays the seller.

A very much more sensible system of showing a great many varieties is that of the demonstration gardens and experimental stations, both stations in which the agricultural authorities try to compare promising new and good old varieties with the aid of objective standards, and such gardens as show- plots arranged by rose-lovers or by societies of growers of gladioli or dahlias. Here the varieties can be judged throughout the growing season in favourable conditions.

THE ANNUAL SELF-FERTILIZING CROPS

The number of plant species now being cultivated by man is really en ormous, w h ile onl y r ela t ive ly f ew a ni ma l s pe cie s a r e k ept un d er domestication. It is possible for the author of a book on animal breeding to take those species almost one by one, and to give the specialist breeders a few hints over and above those which are contained in the general book.

This would be quite out of the question in a book on plant breeding. One way out of the difficulty would be to give one example of each of the three groups—self-fertilized plants, cross-fertilizers and plants that are usually propagated by vegetative means.

These plants in two sections— grains and legumes—giving some hints to the breeders of wheat, barley, oats, sorghum, rice, etc., and going into some details about the breeding of beans, soybeans, cow-peas, ground-nuts, peas, etc., separately. But, after all, the self-fertilizing annuals have so much in common from the point of view of plant-breeding methods.

Self-fertilization causes isolation. Each plant, as far as reproduction is concerned, is as self-contained and isolated in a crowded field as it would be if grown by itself in a greenhouse in a city. The inbreeding caused by self- fertilization has several results.

It makes all heritable variability disappear in every line, it makes every group consist of parallel lines, in which each plant descends from only one parent plant, so that the usual complicated network of ancestry is simplified into a system of isolated parallel lines. Where inherited variability exists, such as happens after a deliberate artificial cross or when accidental cross- fertilization occurs, it will automatically disappear.

That is to say, it will disappear in so far as each separate line is concerned, for Mendelian segregation will make the lines, descending from any hybrid heterozygous for a great many genes, diverse in their hereditary make-up.

The result is that a field of self-fertilizing annuals that has not been grown from just one pure line by a plant breeder always consists of a mixture of a number of genetically different lines. Even if accidental cross-fertilization or irregularities in crossing-over (mutation) are very rare, they do occur, and cause this state of things.

This genetic variability, however, is wholly different from what we find in other plants and animals. We are dealing with mixtures of pure lines, and almost every plant we take from any field is pure, homozygous for all its genes, and will, if its seeds are sown, give a very homogeneous descendance. From all this it follows that enormous progress can often be made in all self- fertilized plants by simply picking out a great number of promising-looking individual plants and comparing their descendance.

Many valuable widely grown wheats, tobaccos, peanuts and beans have been just found. They "just growed ", like Topsy.

The first step in plant-breeding in the self-fertilizing plants must always be to look for those ready-made pure lines. The next is to find whether a mixture of two of the very best pure lines can be discovered which will give us still higher production (10 per cent, higher production from a mixture of two strains as compared with that of the best one is quite common).

We have then to produce some more genetic variability by deliberate cross-breeding. This generally means that we try to combine the good qualities of two different strains in one new one.

Here there are two wholly different schools of thought, and two wholly different methods. One of them is to analyse second-generation plants, and to find those individuals that do combine, let us say, the disease resistance of line A with the large quantity of big seeds of line B. This is a method in- vented by the experimental geneticists who are trying to combine practical plant breeding with purely scientific gene- analysis. As a theoretical geneticist, interested in genes and their action.

It is great fun playing with a hundred different hybrid lots in one species, growing a few dozen second-generation plants in each lot and tabulating the results. But this is neither sound genetics nor sound plant breeding. We are dealing with so many unanalysed and unanalysable genes which all affect production and quality at the same time, that the chance of finding something really worth while in a few hundred plants is simply not good enough.

The other method is based upon the realization that hundreds of genes are commonly involved, and upon the fact that purification—reduction of the potential variability—is rapid and automatic in this material. It consists of growing the greatest possible number of seeds from first-generation hybrid plants. We can then proceed in two different ways.

One is an analytical method, which means that we are seeking for the very best-looking plants (for instance, by weighing their seeds), and progeny testing those plants by sowing a large number of seeds from each, in order to find the most profitable families. The other method is what Baur and Nilsson-Ehle called a "Ramsch" method.

This consists simply of growing the descendance of a hybrid in a vast mixture, refraining from any analysis, and continuing this hybrid lot as a miscellaneous collection for five or six generations, sowing as much of the mixed seed of each harvest as we have room for. If we do this, natural selection will weed out most of the unproductive and undesirable types, such as those plants t1 at ripen after harvesting time, all weaklings and all plants susceptible to diseases and pests. After five or six years that lot will consist of a mixture of pure lines, every line different from every other, with the successful lines represented by more individuals than the unsuccessful ones. After this we will treat the mixture just as if we were dealing with any other mixture, growing separate rows of beds each from one likely-looking plant.

This is a very sound method, which saves a great deal of un- necessary labour and time. It is extremely difficult to find just how much better this method is as compared with that of analysis in every generation, starting from the second generation. An analytical method could be made to work very well, provided we did all we could to make the number of plants of the second and third generations exceedingly large (in the cereals we can get almost as many seeds as in tobacco by splitting up the hybrid plant repeatedly before it starts to grow any stalks).

In actual practice there are only a few crosses that really produce anything worth while, and the general practice seems to be to make a great many attempts and grow a large number of first-generation hybrids. This renders it imperative to cut down on the number of plants per experiment. It is a vicious circle. This is all very wrong and wasteful. One cross every year and work with it, both analytically and by means of a "Ramsch ", doing the final selection after six years for one lot in every season, than waste my time castrating flowers and producing numerous hybrids that would have no really good chance of showing what they could do.

In all the self-fertilizing annuals we must be continually on the look-out for qualities that may be of special merit. Mostly of adaptations to special conditions. In wheaa on irrigated ground, resistance to drowning may be as important as disease resistance or quality of the grain. Conversely, rice could be grown much more extensively if we would take the trouble to do some plant-breeding work on dry-land rices, which abound in the mountain villages of all the tropical Asian islands.

In ground-nuts we should look for lines that are specially adapted to mechanical harvesting, as well as for lines adapted to the production of a first-quality product in the hands of tens of thousands of small cultivators. In all the plants of this group it is possible not only to improve the yield and the quality, but also to extend the area in which profitable production is possible. Tobacco, rice, soybeans and cow-peas are just as important in this respect as wheat, barley and oats.

Plant breeding with self-fertilizing plants is comparatively easy. It is even possible for the real amateur plant breeder to do some extremely interesting and useful work with some of the less well-known plants of this group—let us say with beans or cow-peas. Such work needs room, but very little special technical knowledge or talent for higher arithmetic. Frequently a few plants will stand out from an otherwise quite even field of barley, beans or sorghum.

Such "high birds "—climbing plants in bush beans, high, scmiawned plants in wheat—are often accidental hybrids. If we sow a few of their seeds, we can see immediately whether they are hybrids or admixtures, for hybrids will have very variable offspring. If then we sow all the seeds from such hybrids, we can start a series of experiments that may easily give us something worth while, even when the father of the hybrid is unknown.

HYBRIDING PLANTS

The majority of named clones of fruit-trees, potatoes, strawberries, have been obtained by cross-breeding. This means that somebody has crossed two different kinds of plants and has found the new valuable clone among the hybrids. When we do this cross-breeding work we almost always obtain an enormously variable group, and this shows that the plants used in the cross were heterozygous for a very large number of genes. This is not surprising if we remember that those parents themselves have also been obtained by cross-breeding.

In work with plants grown from seeds, obtaining a hybrid is regarded as only the first step in our attempts to produce something new and very good. W e know tha t we c an neve r ex pect to obta in the full as sortmen t of recombination's of the genes that went into the cross before we have seen the second generation, grown from the hybrids. Why is it that the breeders work in such a different way when they are out to improve their roses, apples or poplars?

In some cases this must be due to the fact that it frequently takes a long time for the hybrid plants to come to maturity. Another cause for this difference in method is certainly found in the great variability in the first hybrid generation, due to heterozygosity of the material.

Whatever the reason for the fact, it is easy to see that the plant breeders who are continually making new hybrids, and then look no farther than the first-generation hybrids, must very often discard hybrid plants that have no value in them- selves, but that would be potential parents of excellent new plants of the second generation. Even if many of the valuable qualities in this material are dominant, it stands to reason that some of the most valuable characters in every group are recessive; and we have not a very good chance of obtaining a full sample of the potential variability that results from a certain cross if we look no farther than the first hybrid generation.

We might even say, without grossly exaggerating, that most named clones in such plants are the result of a cross that has been made with a definite object which has not been attained by the breeder. When we cross two different plants with the object of combining some good quality of one of them with those of the second plant, our work is only half done when we have obtained the hybrids.

To a certain extent we might say that the qualities of the first-generation hybrids are not very important—we must simply regard them as the parents of the variable second generation. Now, in certain groups of plants self-incompatibility is the rule, so that it is impossible to obtain a self-fertilized second generation. In that case we need more than one first- generation hybrid to produce a second generation. But this self-sterility is the exception rather than the rule.

When in plants of this group we sow a great many seeds obtained by self-fertilization, or, by mating together hybrids from the same cross, it is quite possible that we actually realise the hope which set some other breeder working to obtain those hybrids.

Very often we find that some roses, ornamental shrubs or potatoes are setting seed spontaneously. It is frequently asked whether sowing this seed gives the breeder any chance of obtaining something worth while.

To answer this question we must remember that sowing seed with the object of obtaining a plant with good qualities—a valuable novelty—is always something of a lottery.

It is certainly not true that our chance of drawing a winner is made very much better when the seed sown is obtained by crossing two different sorts, unless again we follow through the cross by growing an extensive second generation.

As breeders generally are not in the habit of doing this, almost all the commercial clones have been obtained as hybrids, so that it will very often be worth our while to follow up their descendents.

If a plant breeder, working with vegetatively reproducing plants, has the habit of sowing only cross-bred seeds, he is certain to ascribe his success to this crossing. But we must realise that in so far as valuable new recessive qualities are concerned, we only find those among the hybrids because both parents were heterozygous. Groups in which cross-breeding has always been the rule— apples, lilac, roses, potatoes—sowing inbred seeds on an extensive scale is a method very much worth while.

In some plants, as in the potato, many valuable clones have been obtained from sowing "spontaneously" formed (and presumably inbred) seeds. The method can be especially recommended in those cases where, for some technical reason, such as minute size of the flowers, difficulties in castration, etc., it is extremely difficult to obtain cross-bred seeds. Sugar-cane is a case in point.

F1- Hybrids

In the autogamous plants where genetic purity, once obtained, is easily conserved, it is worth while concentrating our efforts to produce pure- breeding superior lines. But in other cases this may be so difficult that other working methods are preferable. One of those methods is to produce first- generation hybrids for direct use.

In such plants as maize or beets we often find that groups produced by crossing two distinct lines are superior in some way—for instance in yield— to the lines used in the cross. We must begin by giving an explanation of this phenomenon. If we could make a complete genetic analysis of a variable family of plants, we should see that in respect to a very large number of genes the group was pure.

All the individuals of such a group are alike in respect to a great number of genes, for which every individual is homozygous. On the other hand, there are other genes in regard to which such a line is impure, and when we would inbreed in this line, a number of offspring would be produced, that would differ in those genes from the majority. In a variable lot we should find that to a certain extent plants will differ in the set of genes for which they are either pure or impure. The same is true when we are com- paring separate lines.

When we make a cross between two different plants, and especially when we cross two plants of different lines, the result will be that the group of hybrids will be much more uniform than the descendants of either line. The explanation is very simple. The hybrids will inherit all the genes that are common to both parents. In addition to this, they will inherit all the genes for which only one of the parents was homozygous. If one parent had the composition AaBB and the other was AABb, the hybrids will all inherit both A and B.

The result will be that the hybrids are more likely to have valuable dominant qualities than inbred offspring of either line. Where dominance exists, however, and where there is no appreciable difference between homozygous and heterozygous plants (AA and Aa), individual selection will not quickly bring about purity. In other words, it is likely that good individual plants are still heterozygous for valuable genes, whose presence is desired for good quality.

We can understand why, at least in certain combinations of lines, first-generation hybrid lots of plants are superior in quality to and more uniformly good than, pure-bred lots from the same material. In other words, we could say that the homozygosity of one plant for certain genes will cover up the impurity of another plant in respect to those same genes.

We can now understand why hybrid lots are frequently better than inbred lots, and why we find this superiority of the hybrids especially in certain combinations of individuals and of strains. The simplest instances are those in which we combine two pure inbred lines, or even two clones, to produce the commercial seed. Some of the azalea specialists have found combinations of species that, when crossed, give hybrid seed from which beautiful large-flowered hybrid plants can be grown.

A second example is that of the Japanese hybrid egg-plants, and that of hybrid tomato seed. In Java a few tobacco estates have frequently used hybrid seed produced from a cross between quite different lines.

This practice of using first- generation hybrids is commercially possible only if producing such seed is not excessively difficult or expensive. In the cases enumerated, each fertilization gives us a very large number of seeds.

The best-known example of hybrid seed for the commercial production of first-generation hybrid plants is that of maize. Here it will suffice to say that it was found with maize that very often the first-generation hybrid plants produced by crossing different breeds gave a very high yield. The same was found true for hybrids between different inbred lines in one variety.

Starting from small-scale genetical experiments, a very important hybrid-corn seed industry has grown up, first in the U.S.A., and later in other maize-growing countries.

In some cultivated plants (but not in others) inbreeding will have an adverse effect on vigour; and, conversely, crossing between different lines of plants will often give us remarkably vigourous hybrid offspring. Both from a purely scientific standpoint and from a practical one, this subject is of the greatest interest. The facts have been well known for a long time. Darwin made numerous inbreeding experiments, and found that some plants were much more subject to degeneration as a result of it than others.

Man y different theorie s have been proposed to ex plain both th e degeneration due to inbreeding and the great vigour due to crossing. The pre-Mendelian explanations do not concern us much. One of the chief difficulties in the way of a simple genetic explanation has certainly been the old conception of genes as determinants, each calling forth a special discrete unit character.

So long as most geneticists had this conception of genes and of their action, the mere sampling effect of Mendelian segregation of genes and the addition of genes by cross- breeding were thought to be insufficient to account either for the degeneration caused by inbreeding or for the very great special vigour of cross-bred individuals.

It is necessary to go into the different theories that have been proposed and have then been abandoned. The hypothesis that, to my mind, best explains all the facts.

If we look for evidence of loss of vigour by inbreeding, we are first struck by the fact that certain classes of plants are apparently exceptions to the general rule. In the habitually self- fertilized plants, such as tobacco, wheat, barley, and garden beans, the strictest kind of inbreeding—self- fertilization— does not lower the vigour or the yield.

On the other hand, in all the allogamous plants that have been tried—in beets, in maize, in rye—inbreeding brings along a rapid decline in almost all the desirable qualities; and in the same plants, cross-breeding such deteriorated stocks immediately restores full vigour. Now, if we start cross-breeding in the habitually self-fertilizing plants, we often find that the first-generation hybrids are extremely vigourous and productive, and if this happens we notice a very large general reduction in vigour in subsequent inbred generations.

This shows that we must not look for a mysterious special immunity to degeneration in some plants, but that resistance to inbreeding degeneration simply spells homozygosity.

In full accord with this is the circumstance that in animals the facts run wholly parallel to what we find in plants. Those groups in which cross- breeding has been the rule show a decline in quality after inbreeding, and those groups that have been regularly inbred—such as fancy pigeons—are inbreeding- resistant. Again, just as in plants, it has been possible both with chickens and with ducks to produce great purity by a system of inbreeding, while we retain full fertility and vigour as a result of strict selection for those very points.

Of course, the examples of the very fertile and vigourous highly inbred wheats and fowls show that the vigour we see in hybrid stock is not in any way due to the fact that the hybrids are heterozygous in respect to a great many genes, but only to the fact that they possess a favourable combination of genes. In theory it might be possible, by careful work, to make a group of plants (or animals) that would be homozygous in respect to that ideal combination of co-operating genes.

As a matter of fact, that the circum- stance that we do not attain this desirable end is most often due to faulty principles of selection. Essential mistakes are frequently made that render it almost impossible to attain the much-desired results. To understand the nature of inbreeding-degeneration we must remember that dominant good qualities, as far as the effects of selection are concerned, are in a class apart from the recessive virtues. If in a variable group some individuals have some unwanted dominant quality, it is quite simple to eliminate that dominant fault from the material.

If we prefer white maize to yellow, it is quite simple to produce a pure-breeding, white- seeded maize. But the reverse—to produce a pure-breeding, yellow-seeded maize from the same material—is much less simple, as some of the yellow-seeded plants chosen may again be heterozygous.

What is true for one monofactorial difference is true for all of them. If we subject a group of plants to individual selection, or even to some sort of selection by means of progeny tests (e.g., ear to row in maize), and we do not have recourse to inbreeding in order to purify the genotype, the result will be that the superior strains of beets or maize we have built up in this way will be chockful of individuals heterozygous for very important genetic factors.

As soon as we start inbreeding, we reduce the potential variability of such material. In every succeeding generation from self-fertilization the number of genes for which the material was still heterozygous is halved. In every generation the plants become purer and purer.

Let us take the case of just one gene, A. If we inbreed a heterozygote, Aa, 25 per cent, of its descendants will be AA9 25 per cent, will be aa and 50 per cent, will again be heterozygous, Aa. Now, if A is an important gene, the AA and Aa individuals will be somewhat better than the aa's.

If we we re conc erned with th is ge ne only, we could, by merely eliminating the aa plants, work towards better plants, and the proportion of pure-breeding AA plants would improve. But now we must remember that we are concerned with a multitude of genes, for all of which our original material may have been heterozygous.

If we were dealing with only three separate genes, A, B and C, we might retain only the 75 per cent. A plants (AA + Aa) and of these only the three-fourth B plants (AB -f Bb), and of these only the plants with C (again three-fourths). But we must remember that the fraction three-fourth power n becomes rapidly smaller and smaller with an increase of n (the number of desirable genes for which the stock is impure).

For this reason even strict selection for vigour while inbreeding will not prevent a loss of desirable genes from the material in an inbred series. And, of course, blind inbreeding, without any selection, will very rapidly bring down the proportion of plants with dominant desirable qualities, when the material was heterozygous for many genes that helped to build up those desirable dominants.

The mutual independence of genes scattered over the chromosomes of our plants will make it a matter of chance for which desirable genes each inbred line will become homozygous, and which ones will be dropped from the line. It is certain that if we start a number of parallel inbred lines—let us say in beets or in maize—by starting every generation from just one single plant, those lines will become homozygous each for its own genotype.

But the genotype of every line will become different from that of every other line. If we are lucky, and are running a good many lines, we may in the end still possess all the valuable genes that are responsible for the good qualities of superior plants. This does not mean that in one line we may have fixed the plants in respect to one desirable dominant virtue, and that in another we may find another separate good quality.

For we must realise that such good qualities do not depend each on a separate gene, but rather on the correct combination of genes. To a certain extent it is possible that we shall find good qualities in some of our lines— disease resistance in one, good quality of the product in another one. But if the foundation stock was highly cross-bred we shall seldom find highly inbred lines of sufficient all-round quality to make it worth while growing them commercially.

What happens when we cross-breed such inbred lines? As a general rule the vigour and all-round quality of hybrids between two such lines are very much better than those of the lines themselves. Sometimes, but not always, such hybrids are as vigourous and as good as the original material before we started our inbreeding.

The genetic explanation of this can only be that this is a result of the genetic difference between those lines. The loss of quality by an actual loss of genes partly responsible for desirable dominant qualities. Especially in blind inbreeding, chance determines which of such genes will be eliminated in a given line, and which will be consolidated in the line (for which the line would become pure).

If we think of the good quality and good vigour of the original strain as partly caused by the right combination of a great many genes, and of the inferior quality of inbred lines as resulting from incomplete sets of such genes, we can see that the hybrids, which inherit a combination of the genes remaining in inbred strain X with those for which strain Y happens to be homozygous, will be superior to both those strains.

In this superior quality of the hybrid generation between inbred lines we are concerned with the genotype, with the combination of useful genes present in the plants, and our present-day explanation is that it is this good combination of genes only which confers good quality upon those plants, and that the fact that they are heterozygous for many of them does not help. It would be very desirable if we could devise means for making a strain of plants that would be homozygous for all those important genes.

HYBRIDS OF MAIZE AND BEETS

An Example of Utilizing "Hybrid Vigour"

From a small beginning in the experimental plots of the geneticists (G. Shull, East and Jones), the work with hybrid vigour in maize has grown to enormous proportions. The greater part of the corn harvest in the U.S. A. is now produced from hybrid seed, and the yield per acre has been much improved thereby. In the older well-known maize varieties that formerly were always grown from open fertilized plants the potential variability was very great. Many plants are heterozygous for very many different genes, and the best vigour and the best yield are found in those plants which happen to possess the most complete combination of those genes.

If we start inbreeding in series of lines, so that in every line we take one plant and force it to pollinate itself, we quickly reduce the potential variability in each of those lines. The plants will be homozygous for many genes for which the original mother-plant was still impure, and, on the other hand, we shall lose some of the useful dominant qualities.

The general quality and vigour of the inbred lines will decline, especially so if we use blind inbreeding without selection in continuing the lines. If, after some generations of inbreeding, we now cross those inbred lines, we shall find that, in some combinations more than in others, the resulting hybrids will be of very good quality. We have explained this by the hypothesis that those hybrids again have the full set of valuable genes. Now, the question may be asked why it should be advantageous to go to all this trouble of purifying those separate lines in order to reconstitute the original set of genes by cross-breeding them. Why should the hybrids between a good combination of those lines be more valuable than plants from the original non-inbred strain? There are at least two reasons for this. In the original lot—in the strain, before it was inbred—the potential variability was very high.

A large proportion of the plants—probably all of them—were heterozygous for several of the genes that help to make for quality. The germ-cells produced in the field were of the most diverse kinds, and, as a result, a certain percentage of the plants of the next generation always had undesirable recessive traits. Maize and sugar- beets are full of the most diverse undesirable recessive characters, that crop up in every generation.

The inbred strains may not have a full complement of the desirable genes, but in any case they are homozygous for those they do carry. If, for this reason, we are able to find a set of two inbred lines that "click" because together they happen to have a full complement of the desirable genes, the result will be that the hybrids grown from this combination will all carry the whole set of those genes. For this reason the hybrid lot will be much less variable than the original open-fertilized variety. In other words, the method of establishing many different inbred lines is a method of cutting down the heterozygosity in the material. For practical reasons, this reduction of the variability is not attempted in one single series.

The breeders know that it would be impracticable with our present methods to produce just one inbred strain, homozygous for all the desirable genes, so they do the work in parallel series, fixing the genotype of several lines by inbreeding, and hoping that after several generations of this they will discover a few inbred lines whose combination will give them the desired hybrid seed that will produce vigourous high-yielding plants.

It is obvious that the quality of such first-generation hybrids depends on the combination of genes present, but that the plants themselves are again heterozygous for all those genes. Those hybrid plants are meant to be harvested for their produce—seed from such hybrid lots would give as variable a supply of plants as the original strain, or almost so. From the point of view of the seed firms this circumstance is rather happy; the users of the seed are satisfied, but they have to come back next year for a fresh lot of seed.

The fact that most inbred strains of maize and beets are of very low vigour and production makes the production of such hybrid seed expensive. In many maize-strains the cobs are miserable nubbins, the seeds are small and irregular in shape and size, and poorly adapted to the use of sowing by machinery. There are several ways out of this difficulty. Obviously the best way is to produce at least one line that will be a fairly good producer of seed.

The methods of blind inbreeding of the first years have been improved, and it is usual nowadays to combine inbreeding with selection. Sometimes this means straight selection in the inbred line, sometimes the selection is used to fit the inbred strain for its ultimate use—cross-breeding, We can do this by testing the pollen of several plants in cross- breeding (top-crossing) tests, and so choose the sub-line which is retaining most of the valuable genes.

Another method which is often used is that of the triple cross, in which the hybrid plants that furnish a lot of good- looking and very even seed are fertilized by pollen of yet a third line, that has been carefully chosen for this very purpose. If the ultimate product must be very uniform—such as in sweet- corn that is grown for canning, and that must be prepared by machinery— the use of one single cross between two inbred lines is indicated. The use of hybrid plants as parents of the commercial seed gives too variable a crop. In the ordinary use of maize some variability can be permitted, provided the average yield is good. In certain circumstances inbred strains can be produced that are valuable for "top-crossing ".

Such strains must be pure for a great many important genes, and they must be good pollen producers. The usual practice in producing hybrid seed for sale is to grow the mother-plants in a field, and grow the pollen parents throughout that field in every third or fourth row. The mother-plants are then carefully detasselled, and the field produces hybrid seed from these detasselled plants and inbred seed from the pollen-bearers.

Nowadays much of the commercial maize seed is grown by the system of "double-crosses ". In this system four different inbreds are used, and to grow the commercial seed, hybrids A × B are crossed with hybrids of two other strains, C × D. This, of course, makes the production of commercial seed much cheaper. It is doubtful if from the standpoint of the seed-user this system is as good as the triple cross.

We must realise that the hybrid plants of the first cross are heterozygous in respect to all the genes in which the parents differ. A uniform descendance from such plants can be expected only when the pollen used is from plants homozygous for almost all those genes; and we cannot expect that the pollen of another hybrid is in that condition.

It is certain that even commercial double-cross seed, when well-chosen inbreds are used, gives a better harvest than ordinary mass-selected stock of the same variety. But the results are never as good (and could not be expected to be as good) as those obtained from single-crossed seed. A certain extent smart advertising plays a role in this subject of the hybrid-corn seed production.

In the catalogues of the seed firms we often see pictures of "typical" ears of inbred strains, of the first hybrids, and of the "double-cross ". The enormous improvement in the first hybrids over the inbred strains, but they always show the largest and finest cob as the result of the double cross. Such pictures must be taken with a grain of salt!

If we look at the production of seed maize from the commercial angle, we must realise that the process of inbreeding those sub-strains costs money. During the first generations many plants have to be enclosed in paper bags, and to be pollinated by hand.

But, on the other hand, when the difficult and painstaking labour of purifying and testing the inbreds is done, those inbred strains can be further multiplied, without any manipulation, in fields, sufficiently far removed from other fields of corn. In fact, once we have found the two suitable inbred strains that "click", or the three necessary for a triple cross, we can reduce our selection to the very minimum, for once such strains are homozygous for the important genes, they remain in that condition, and all we have to do is to make the final cross to produce the seed-crop for sale.

The farmers, especially in the U.S.A., have become "hybrid- corn conscious ". They have experienced on their own farm how much more vigourous are the crops that can be obtained from the hybrid seed. The idea of the superiority of cross-bred seed has taken firm root, the name "double-cross" has more advertising value than the facts warrant. The only yield figures ever seen in which "double- cross" plants gave a (slightly) better harvest than their single- cross parents were taken from a cross-breeding experiment in which four different varieties were used.

With beets, especially sugar-beet, the commercial houses use a system of seed production that has great similarities to that of the hybrid maize. They generally use a system of mild inbreeding in isolated lines. Inbreeding is much less easy here, and whole plants must be enclosed in very well- constructed and ventilated cages. Next they try combinations of their selected strains, and finally they grow their commercial seed from a field in which two strains are planted in a mixture.

Until recently there has been no unfailing method of producing 100 per cent, hybrid seed, as a certain percentage of the seed must come from self- fertilization or from the inter- crossing between different plants of the same line. In practice, however, we find that the vast majority of the beets in a commercial field are first-generation hybrids.

If we deliberately cross two strains of beets, in which we can easily recognize the hybrids—by their leaf-colour or by their shape—we find that, with very few exceptions, the field contains only hybrid plants. The cause of this must be sought in the great vigour of the first-generation hybrids. In sowing beets many more seeds are sown than will ultimately grow up into mature beets.

Soon after the seedlings are up, the field is thinned, the supernumerary plantlets are hoed out. As this is done by hand, the man who does the thinning always favours the strongest, most vigourous plantlets. Even if we did not consciously select the favourable combinations between lines to be combined for seed-growing, the percentage of mature beets that have been grown from cross-fertilized seed is always very high.

Further Improvement in Maize and Beets

We saw that inbreeding will reduce the potential variability of the families subjected to it, that those lines will become homozygous for some of the valuable genes and that they will lose some of the others. One question that is of the greatest theoretical, but especially practical interest, is what we could do to influence this process of purification in such a way that each line would become pure for the maximum number of useful genes.

For discussing this, we must again look at this process of automatic reduction of the variability during inbreeding. We start with a plant of maize or of sugar-beet that is heterozygous for a great number of valuable genes— genes that we would like to have in every plant. The ordinary selection work ensures that such genes remain present in the stock, even if ordinary mass selection does not quickly ensure purity in respect of those genes.

It does not matter whether we know or whether we do not know all such genes separately in their action upon the end result. But we do know that a plant heterozygous for a gene A (Aa)y on being self-fertilized, will give us 25 per cent, homozygous A A plants, 50 per cent Aa's and 25 per cent aa's (which last will tend to be discarded).

The chance that we will hit upon a plant homozygous for one of those genes is one in four; if we are dealing with five genes, we have only one chance in eight hundred of finding a descendant homozygous for all of them— in fact, the chance is one in 4" (n being the number of genes involved) that we find a plant homozygous for all the genes concerned. In a limited number of inbred plants all we can hope is that we conserve some of those useful genes in the heterozygotes.

In the work with hybrids between lines we help ourselves by continuing a number of parallel inbred lines, hoping that in the end we shall have two such lines which between them carry the whole series of important genes in homozygous condition. It is evident that (theoretically) we should be able to produce inbred lines of maize or sugar-beets that were as productive as some of the best strains we now possess, and that would be homozygous and breed true.

In fact, if the number of inbred plants from a very good parent plant were only large enough, we should be able to find such a plant homozygous for all those genes for which the mother was still impure, in the very fast generation. In practice we can never know whether we would have to sow two thousand or twenty thousand plants to find that ideal one. But one thing is certain: that if we sow a few dozen plants from self-fertilized seed we shall undoubtedly have no chance whatever!

The simplest genetic calculations show that in this selection work in such plants as maize and beets the usual number of individual plants grown per generation is far and away too small! Not only should we combine selection with inbreeding, we should see to it that the methods used gave us an adequate chance of finding what we want to select.

Of course a great difficulty lies in the fact that it is practically impossible to appraise the genetic quality of one plant by inspecting its individual quality. This difficulty is, of course, greater in beets than in maize. Nevertheless, it would pay well to start inbreeding series on a very wide front, sowing all the seeds from an excellent plant, and testing a number of very excellent daughter-plants by sowing their inbred progeny.

The circumstance that inbred series almost always deteriorate rapidly in vigour and yield does not prove a thing, so long as the numbers we work with are so inadequate. A technical difficulty, of course, is that in maize it is not possible to pick out the individually best plants before they have ripened their seeds. When we must pollinate them we do not know their quality. Beets give much less trouble in this respect.

A very great difficulty with which plant breeders have to contend in the allogamous plants, lies in the large number of genes in respect to which our material is variable. Inbreeding will decrease this potential variability, but, as we have already seen, it is extremely difficult to direct the purification in such a way that the inbred lines will be of good commercial value. There is one way out of this difficulty. We know that the inbred lines are highly homozygous.

If th ey a re not good en ough a s the y a re, we kn ow neverth eless tha t such lines a re homoz ygous for m a ny va lua ble gen es. It is well worth while trying to improve such pure inbred lines, and this is being widely tried in America to-day. We can do this in two ways: by crossing to an oth er line, followed by an other series of in breedin g, or by the us e of repeated back-cross es. A cross with another suitable line will give us the genes lacking in our inbred lines. But the hybrids will be highly impure, and we must try to reduce this impurity while profiting by the homozygosity of the inbred line. This we can best do by a combination of repeated back-crosses to that line, and selection for vigour and yield.

In the existence of good inbred lines in maize, that can be propagated separately, we have an excellent and sure method of producing first-class commercial seed for sale. In the beets we are far from this. But in the beet there are certain possibilities that should be worth trying out. In the first place, there is male sterility. Certain families of beets have been found that have no functioning pollen. maternal lines, and it seems to be due to a peculiarity of the cytoplasm (protoplasm outside the nucleus). By cross-breeding it can be brought into any strain of beets. This gives us a possibility of producing on those plants seed that is guaranteed cross-bred, by having such plants flower in a field with other plants.

The other possibility in beets is the use of vegetative reproduction. Let us suppose that we possess two plants of sugar- beet that happen to fit together so well that the hybrid seeds produced from their combination give us beets of high purity and sugar content and with an excellent yield. Now our difficulty would be to produce two different groups of beets that would have the same constitution as those two individual beets.

This would be extremely difficult and tedious work. On the other hand, if we had a very good method of rapid vegetative reproduction, all we would have to do would be to grow tens of thousands of beets of clone X and as many of clone T, and then to make them flower in a mixed field in the ordinary way. There exist a few methods of vegetative reproduction in beets. Some of these methods, involving making the beets produce shoots and reproducing new beets from cuttings from those shoots, work fairly well.

If the methods can be perfected, they could lead to an avoidance of the very costly methods of selection now in use.

The Choice of Plant-breeding Material

The advantage of having at one's disposal a large collection of plant varieties, both plants of commercial importance and imported plants of no immediate practical importance, is enormous. A difficult question for the plant breeder is always what crosses to make, what plants to combine with the greatest hope of ultimate success.

In the breeding of such plants as potatoes, raspberries and roses, certain clones have the reputation of being profitable parents of hybrid lots. The Early Rose potato, the Lloyd George raspberry, the Chilean strawberry.

In some special instances the choice is obvious. When our present assortment of flaxes or potatoes or roses happens to be deficient in some such quality as disease resistance or winter- hardiness, we are on the look-out for plants that will show those desirable qualities to the greatest extent. In other words, we shall, when choosing the parents of our future hybrids, try to find material that promises the possibility of a combination of the very good points of both.

This is obvious. Often crosses that seemed very little promising have given excellent results. It would seem that here no definite rule could be indicated. It is always worth while trying crosses with new material, with newly imported species—in fact with material which has not been used previously. This is especially so when, as in horti- culture, everybody is on the look-out for something new.

Another rule that could be given would be that when we make very wide crosses we cannot hope to find commercially valuable strains in the second hybrid generation, unless we grow extremely large numbers of plants in that generation. When we make a very wide cross to incorporate some special quality into the existent stocks, the best thing we can do is to use the hybrids in back-cross series.

We can then strive to retain just one special quality of the new species and incorporate it into the general set of genes with which we are familiar in our older stock.

In several instances very wide crosses have been successfully used, and sometimes plants have been used in cross-breeding that in themselves showed very little promise. The inedible cattle gourd that Orton crossed with his water- melons, of some of the potato crosses with Andean wild species (Salaman) and of the back-crosses to hybrids with wild cane from which all our modern resistant varieties of sugar-cane have been derived.

It is here—in the choosing of plant material for breeding work—that cytological studies and chromosomal counts are of the greatest practical value.

Methods of Evaluation and Comparison

The aim of every plant breeder is to produce a group that will be worth a place in the assortment of cultivated plants. In some cases the new plant will have only ornamental value, in others it will, for any of a dozen reasons, be better adapted to agriculture than other similar ones. It is easy enough in all classes of plants to produce novelties, to produce variability. The experimental grounds of thc plant breeder would soon be cluttered up with novelties if he did not keep on weeding out the comparatively worthless. After producing our novel plants, we must judge them.

Some qualities are much easier to judge than others. It is comparatively easy to weed out a lot of seedling dahlias and to pick out a few that may be worth keeping on for further testing. In the selection of our agricultural plants, in cereals, in beets, in maize, the assortment of valuable kinds is already so great that only relatively few of the newer selections are worth keeping. In the vegetatively propagated plants this work of testing novelties is comparatively easy.

In the first place, we are not bothered by considerations of purity— once we have produced a very good specimen, we can just multiply it with all its properties intact. In the second place, we can easily, for purposes of testing and comparison, produce as many individual plants of the novelty as we require.

When the differences are relatively small—as in cereals and beets, where the novel plant produced must compete with excellent old stock that has been made by a process of selection and comparison during centuries—a correct system of evaluation becomes of very great importance. What we really want to test is the inherited composition of the material.

This may be difficult, as the final quality of the plants to a large extent depends on favourable or unfavourable conditions under which they grew up. It will perhaps be as well to begin with the simplest problems. In the production of novelties in such flowers as roses or dahlias the value of a new variety depends entirely on the appeal it makes to the gardening public.

The man who is proud to have found a novel dahlia seedling or a new and strikingly beautiful rose sends it to a show and hopes that it will take the judge's eye and win him the much- "prized ribbon or certificate. But of course the insider knows that many highly commended novelties will never be heard of again, or that they will drop out of cultivation in a few years.

There is a difference between the fact that the grower can bring half a dozen perfect blooms to the show on time, and the question whether that same rose or chrysanthemum will ever be popular in the gardens, or even be worth while growing for the commercial producer of cut flowers. A dahlia that wins the medal may be a very weak grower, or it may be very susceptible to rain.

The strikingly beautiful rose may be very susceptible to mildew, or its flowerin g seas on may be extremely sh ort. If we visit th e grow ers of greenhouse roses, or the gardens of the man who produces cut flowers for the market, we find that the number of varieties that have real commercial value—varieties "with bread in them ", as our gardeners express it—is extremely small. We shall probably find only three or four of the relatively older dahlias worth growing for cut flowers, and it is seldom that a novelty is added to that list.

In testing out agricultural plants, both during the process of selection and after they have finally been introduced, many methods are in use. The direct, simple examination of a number of small plots or rows will show us the qualities of those lots, but this method will only serve to find the groups with definitely inferior quality.

It is not good enough to find exceptional merit, as circumstances may have favoured those qualities we appreciate. The only way to avoid reaching the wrong conclusions is to see to it that environmental conditions do not vitiate our results. The best way to do this is to com- pare the quality of a number of plants with that of other plants that were grown in wholly comparable conditions.

One of the best ways to do this is to grow the plots of our plant between similar plots of an older variety, that can be used as a standard of comparison. The choice of this standard is of great importance, and we will revert to this subject when treating of the breeding of special plant groups. As an example we can here take the evaluation of a newly produced (or of an imported) potato clone, or pure line of wheat or barley, In a case like this the standard may be some line or clone that is of commercial importance, against which the tested plant will be called upon to compete. If there is sufficient material, it is wise to repeat the comparison—that is to say, we should grow the test plot between standard plots in different parts of the field, or preferably in different fields. To equalize accidental conditions in small parts of the field, oblong plots are generally better than square ones. If we are testing a number of experimental lots, we can grow them side by side, and interpolate standard plots to be used for purposes of comparison. The experimental garden will need an efficient system of book- keeping. We must note down anything of value about the experimental and the standard plots, date of maturity, disease resistance, quality, yield.

We can calculate average yield for the sum of the plots of one kind, and then compare those figures with the average computed yield of the standard plots or with the average yield of the whole field; and a comparison of the deviation of the yields in the different plots with calculated mean errors will give us valuable data about the question in how far the superiority or inferiority of each separate number is significant as an indication of merit.

The practice of plant breeding in different parts of the world. A multitude of miserably small checkerboard fields being weighed at harvest time to p roduc e th e f i gu res for a n of fice sta f f to juggle w ith . A s uc c es sful establishments where the field-work was under the supervision of men who were thoroughly conversant with the material, and where excellent results were obtained with a minimum of calculating machine work.

A method of annotating the results obtained in experimental plots which gives the plant breeder an excellent insight into the merits of his new material is that of graphical registration. This method can be adapted to the most diverse projects; it is very simple, and gives us our results at a glance. The method consists of registering the quality (yield, height, resistance) of each plot, both standard plots and testing plots, on ruled paper.

If we do this, the line that can be drawn through the points of the standard plots will probably rise in one part of the field, and descend in some other part of it, and by drawing a line that joins those points we obtain a basis upon which we can found our judgement of the results obtained in the experimental plots.

If we want to test our experimental material in two or three different qualities, we must construct separate graphs for each of those qualities, or sometimes we can use the same graph, employing different colours for separate qualities (weight, sugar percentage, purity in beets, etc.).

Reading those graphs will now show us the quality of our plant plots, in comparison with the standards, and we can use this relative value as a basis for our selection work. If we compare half a dozen groups in one experiment, and we find that one or two of them will in every duplication remain well below the standard in quality, this may be a very good reason for discarding them.

On the other hand, if we find one or two numbers that all through the experiment are well above the standard line in quality, we have a very good reason for preferring them.

It may be that in the case of one or two numbers the results are very variable, so that in one part of the field the number is above and in other parts it is below the standard in quality. Then we may well retain this number for another trial in the next season.

The influence of one row of plants on those growing close to it may be considerable, and we must guard against this. If we can grow three rows and use only the middle one for our records, we take a step in the right direction. If we have enough seed to plant six rows, and we examine only the two middle ones, discarding two rows on each side, this is still better.

The greatest difficulty arises in those cases where we are dealing with very few plants of each number; or, worse, where, as in the second generation of a cross, every plant is probably different from every other. It then becomes very important to equalize conditions for each plant as far as this is possible. A very good plan is to grow the plants in rows, and a lterna te those experimental rows with rows of some pure standard variety.

Each plant may then have two diverse neighbours in its own row, but it will touch four other plants that will be of the same kind as four out of six neighbour plants in the case of every other plant in the experiment. This means that to save time we must make a preliminary choice out of the many numbers in those early generations, and it is very important that we shall make the cultural conditions as nearly equal as possible for every plant. This is the place to warn against using apparent correlative qualities in judging the value of our experimental plants. In the breeding of animals even to-day we often see that almost as much attention is paid to correlative qualities as to real merit.

Some of this is certainly due to the fact that in some material real merit can only be judged in one of the two sexes. There is no good reason to give any attention to correlative qualities. In so far as a real correlation exists, selection for real quality will automatically bring along the correlative character, and in as far as we are mistaken (and this will nearly always be so) we are simply led astray by paying attention to apparent correlations. In plant-breeding selection according to qualities believed to be indirect indications of good quality used to be very frequently attempted.

This has gone out of fashion, and we only meet traces of it in very special fields (sugar-cane, apples). In our selection work we should always consider actual practical merit, and we should care- fully study in how far the qualities we can examine, measure and weigh are indications of this real merit.

The Importance of Collections

Most plant-breeding institutes own a collection of plants, sometimes consisting of imported and old varieties, sometimes completed with a collection of wild species. Such collections are of the utmost importance to plant breeding. We must remember that by selection we cannot create new genes, we cannot even remove existing genes that are present in all the plants.

For this reason the possibilities for plant- breeding work are restricted to recombination of the material present in our collection, The importance of a collection of old varieties is sometimes seen in the historical interest, and occasionally old breeds are retained for sentimental reasons. This is wrong, for at any time somebody responsible for the collection may have other ideas; he may be less of a sentimentalist than his predecessors, or he may think the value of all those old varieties is over- rated now that we have so many superior new ones. The truth is that the collections furnish material for serious future plant- breeding work. Every new gene added to the complement of genes in our present material doubles the possibilities of recombinations. Every novel (relative) gene-absence does the same. It is some- times possible to use very important qualities of intrinsically worthless old kinds of plant.

A garden pea, called "Acacia". In this pea each leaf consists of an uninterrupted series of pairs of leaflets like a Robinia leaf, tendrils being wholly absent. The plants therefore cannot attach themselves to supports or neighbouring plants. One of the difficulties in growing field peas in wet summers in some countries lies in the circumstance that the young plants intertwine themselves into veritable mats before the time for hoeing the weeds is past. To give the plants another cultivation in May would mean tearing apart the tendrils; in most years this would be quite out of the question.

For this reason the "Acacia" variety with a good, green, round field pea, and by selecting among the plants of the third and fourth generation succeeded in remaking a green field pea with a good yield and a high resistance to mildew, but without tendrils, a plant which could be given a last hoeing fully three weeks after the ordinary peas in neighbouring fields had grown together inextricably.

When we want to extend the region in which we can grow some group of plants it is often possible to find plant varieties in an extensive collection which carry qualities that are the very thing we want. It is uneconomical to duplicate and triplicate such collections; it is much better to have one huge central "ecole" of cultivated plants than to have a number of incomplete collections in the hands of private plant breeders and plant-breeding firms.

Moreover, keeping up such a collection costs money, and for this reason the collection, or some of its less- important-looking species, always runs the risk of being sup- pressed some day. It is evident that the Government should take over the task of keeping up such plant collections. The American Government has acted upon this idea, and its Plant Importation Bureau has done an enormous amount of very valuable work in this connection.

The keeping of a collection and the importation of foreign plants should go hand in hand. Especially in backward countries a wealth of plants is often present, and we sometimes find a great many curious plants well adapted to special conditions of season and climate. Plants as the Hopi maize, which comes up from almost two feet underground, the pea that buries its pods like a peanut, and the cauliflower-like sugar-cane of some of the islands around Java.

Hull-less oats and hull-less barleys may be out of fashion in some countries just now, nevertheless they carry extremely valuable characteristics, well worth preserving. Think of the extremely hardy peaches and apricots of the higher Himalayan valleys, of blight-free Chinese chestnuts, of the enormous seedless tangerines from Java. Keeping a collection is no work for a geneticist; it needs its own specialists with a very special turn of mind—a zest for systematic work and for collecting. Perhaps philatelists might make excellent keepers for such collections! The chief source of plants in such collections must be importations from newly explored countries. Sometimes, however, it might be worth while to add some of the curious plants that turn up in the descendants of hybrids—plants carrying absolutely novel qualities, new for the group.

Having an important collection of cultivated plants on hand makes it unnecessary to start hunting for suitable material whenever we want to take the amelioration of some plant group in hand. Most plant-breeding firms and plant-breeding institutes keep some sort of a collection, mostly of plants that have or that have had some economic significance. It is of the utmost importance to centralize this work in every country.

❐

9

Integrated Techniques for Overcoming Pre and Post fertilization Barriers

Pollination is the transfer of pollen to the female organs of seed plants. In flowering plants (angiosperms, or "covered seeds"), immature seeds (ovules) are located within carpels. In contrast, nonflowering seed plants have uncovered ovules to which the pollen is transferred, making these "naked seeds" (gymnosperms). Gymnosperms have simpler pollination as all transmit their pollen by wind.

In contrast, angiosperm have a wealth of pollination methods involving many different agents to transfer pollen, including insects (entomophily), birds (ornithophily), bats (chirophily), wind (anemophily), and water (hydrophily).

Attraction of animals usually occurs through a conspicuous floral display, with colour and scent playing important roles. Yellow and blue flowers tend to attract bees, red flowers attract hummingbirds, pink flowers attract butterflies, and white flowers that stand out at night often attract moths and bats. Animal-based pollination is efficient and usually associated with a food reward to assure a continuing relationship.

Frequently, this is a simple symbiotic relationship; however, other plants seem to practice deceit. Trap pollinators may hold insects hostage until a flower is pollinated successfully, when they are released. A most unusual mechanism are orchids that seemingly imitate the form and scent of female wasps, attracting amorous male wasps to mount the flower, inadvertently pollinating it.

Aquatic plants frequently have filamentous pollen, which is more easily captured in water, and may release male flowers that float freely to their attached female counterparts.

When pollen is deposited on the stigma (in angiosperms) or the ovule (in gymnosperms), it germinates, forming a slender pollen tube through a weakened area of the pollen wall. The pollen tube elongates through "tip extension," penetrating between cells of the host parent. Within the pollen tube, two nonmotile sperm cells are ultimately formed and are conveyed through the tube, keeping pace with tip growth.

The pollen tube uses chemotropic signals to determine the final pathway to the egg cell, deep within the ovule. In angiosperms, pollen tubes penetrate the stigma, style, and ovary until they are amid the ovules.

In gymnosperms, pollen germinates directly on the ovule. Pollen tubes enter ovules through a tiny pore called the micropyle and then elongate into the f ema le gam etophyte (c alled th e e mbryo sa c in an giosperm s). In gymnosperms, the pollen tube directly penetrates the egg cell, but in angiosperms, there are sterile cells in the embryo sac, called synergids, that initially receive the sperm.

At this point, one sperm cell is discharged from the pollen tube and fuses, with the egg cell to form the zygote (the immediate fusion product) and subsequent embryo, which will become the off spring plant. In angiosperms, the second sperm fuses with the central cell to form a nutritive endosperm during double fertilization. The endosperm is needed for successful embryo development.

During fertilization the male and female gametes:

1. contact one another,
2. adhere,
3. cells fuse,
4. nuclei fuse.

The act of fertilization triggers embryo development in all plants and endosperm development in angiosperms.

FLOWERS

Flowers are the typically showy reproductive organs of angiosperms (flowering plants). Their diverse blooms generate countless horticultural products; functionally flowers are essential for sexual reproduction. Flowers may occur individually on the plant or together with other flowers forming inflorescences. Their flowers and inflorescences are:

1. highly adapted to the plant's ecology and pollination strategy;
2. highly variable among the flowering plants;
3. critically important to plant identification;
4. subject to their own rich and specialized terminology.

Morphologically and evolutionarily, the flower is regarded as a terminal shoot with variably leaflike lateral appendages. The stem of the flower, known as a pedicel, is connected to the other floral parts via the receptacle. From the outside rim of the receptacle, the first appendages are typically green, leaflike sepals, collectively called the calyx. The next whorl is composed of variously coloured and often complex petals, collectively called the corolla.

The androecium forms the next whorl, formed of stamens, each made of a filament and anther. Anther sacs are filled with pollen at maturity. Anthers release the pollen timed to match the receptivity of the female of either the same flower or others of the same species. The centre of the flower is occupied by the gynoecium, which is formed of one or more carpels.

Carpels may fuse to form a compound pistil or may remain separate and unfused. The carpel is divided into three regions, from the tip: the stigma, a receptive surface for pollen; the style, a specialized region for pollen tube elongation; and the ovary, a location where immature seeds called ovules occur, containing the female gametophyte or embryo sac. Meiosis occurs deep within the anther to produce many haploid male gametophytes (which become pollen) and deep within the ovule to produce the female gametophyte (or embryo sac). The group of angiosperms known as monocotyledons typically have floral organs in multiples of three, whereas dicotyledons have floral organs in multiples of four or five.

Fusion of floral organs often occurs obscuring these parts and increasing the diversity of floral form. Similar organs fuse in the example of the corolla tube (as in the morning glory), but dissimilar organs fuse in the example of apple flowers. Organs may also be modified to perform different functions. Roses have five petals (as do most dicots), but some or all of the numerous stamens form petal-like staminodes, sterile petal-like stamens that give horticultural roses the appearance of having many more petals.

Symmetry of individual flowers may be radial (forming mirror images around the centre), or bilateral (forming mirror images along only one plane). The flower is the focus of considerable metabolic activity for the plant, producing reproductive organs with high energy content. Flowers attract Snapdragons in bloom. Insects and other animal pollinators that visit flowers are often rewarded in return with pollen or nectar to assure this continued relationship. Insects and other animal pollinators to visit the flower.

In return, pollinators are often provided with a food reward of pollen or nectar to assure this continued relationship. After the flower is fertilized, floral organs, including petals, stigma, and style, die back while seeds form inside the carpels. The tissues surrounding the ovary are dramatically modified to become the fruit.

PLANT DEVELOPMENT

Plant development is an umbrella term for a broad spectrum of processes that include: the formation of a complete embryo from a zygote; seed germination; the elaboration of a mature vegetative plant from the embryo; the formation of flowers, fruits, and seeds; and many of the plant's responses to its environment.

Plant development encompasses the growth and differentiation of cells, tissues, organs, and organ systems. Plant development shares many similarities with developmental processes in animals, but the fact that plants are nonmotile, photosynthetic organisms requires certain novel developmental processes in addition to the common ones.

Embryo and Seed Development

Embryogenesis, the formation of a multicellular embryo from a single-celled zygote, is one of the most dramatic and best-characterized aspects of plant development. Four key developmental processes take place during embryogenesis. First, the zygote expresses apical-basal polarity, meaning that the apical and basal ends of the zygote cell differ structurally and biochemically.

When the zygote divides, it typically divides asymmetrically, giving rise to a small apical cell with dense cytoplasm and a large basal cell with watery cytoplasm. Although these two cells have identical nuclei, their fates differ dramatically. The apical cell gives rise to the embryo itself, while the basal cell gives rise to a short-lived structure called a suspensor and the tip of the root system.

The progeny of the apical cell grow. Embryo formation begins with cell division that establishes the apical-basal (top-bottom) axis. Further divisions elaborate on this basic plan, finally forming the cotyledons (seed leaves), as well as the apical meristems of root and shoot. Spherical mass of cells, the globular-stage embryo. Second, differential growth within the globular embryo gives rise to the "heart" stage embryo, the earliest stage when the precursors of cotyledons, root, and stem can be recognized.

This key embryogenic process is called organogenesis. Third, distinctive planes of cell divisions bring about histogenesis, the process by which cells within embryonic cotyledons, root, and stem acquire different shapes, forming the precursors of the plant tissue systems. Last, the apical meristems of the shoot and root systems are formed at the apical and basal ends of the embryo. After an embryo has reached full size, developmental changes continue to occur at the cellular level.

Embryonic cells, particularly those of the cotyledons, begin to synthesize and store the proteins, lipids, and starch that will provide the energy and basic building blocks for germination and seedling growth. Next, the embryo begins to desiccate, sometimes losing up to 80 per cent of its previous water content, and enters a phase of dormancy.

Development and metabolism are arrested in dormant embryos, and seeds containing dormant embryos can survive for many years (sometimes centuries) and withstand extreme temperatures and drought. Plant hormones are important regulators of embryogenesis and seed dormancy.

The hormones auxin, gibberellic acid, and cytokinin all stimulate growth and are present in the embryo during the stages of embryogenesis. As the em bryo mat ures , th ese horm on es a re degrade d an d a bscisic a cid is synthesized by the embryo.

Abscisic acid provides a developmental signal for the embryo to initiate the synthesis of storage compounds and to undergo desiccation. Abscisic acid is present in dormant seeds and is thought to play an important role in maintaining seed dormancy.

GERMINATION AND SEEDLING DEVELOPMENT

Embryo development and metabolism resume upon seed germination. Given the right combination of water availability, temperatures, and light, the desiccated seed begins to take up water and the embryo begins to grow and metabolize again.

Some species have specific requirements for germination; for instance, many temperate zone tree species require several weeks of temperatures of 4 degrees Celsius (39.2 degrees Fahrenheit) or less.

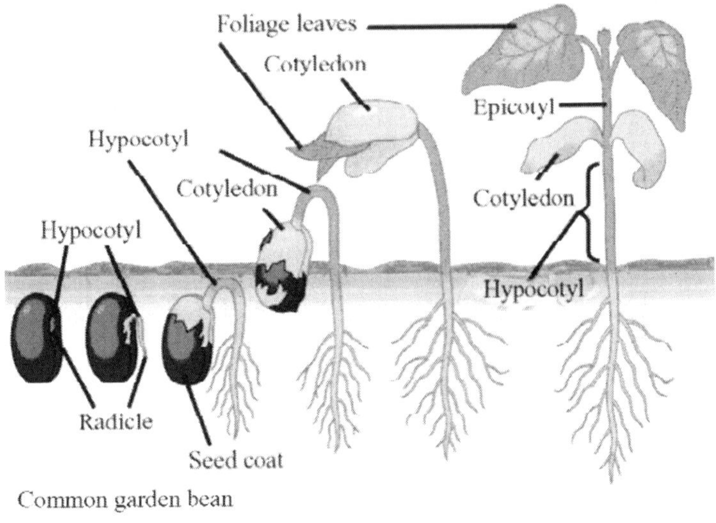

Fig.Germination of Seed

The root is the first portion of the plant to emerge during germination. Growth of the stem behind the cotyledons forms a "hook" that emerges from the soil, followed by emergence of the cotyledons, which begin to photosynthesize to feed further growth. order to germinate. Other species require low levels of light in order to germinate. Once germination is initiated, the embryo follows a typical pattern of development. In many plants, the preformed embryonic root elongates first, forcing its way out of the seed coat and into the soil.

Next, the embryonic stem, usually the part below the attachment of the cotyledons (the hypocotyl), elongates. Once the hypocotyl has carried the cotyledons into the light, they expand, providing a broad surface for photosynthesis. Environmental factors and their translation into hormonal signals are important for seedling development.

For instance, germination in the dark results in developmental events that help the seedling push its way through the soil into the light. The hypocotyl elongates quickly and maintains a "hook" near its tip that protects the cotyledons and shoot apical meristem region. Cotyledon expansion is suppressed so that they are not damaged as they are pulled through the soil. In contrast, if the same seeds germinate in the light, the hypocotyl hardly elongates at all and does not form a hook, while the cotyledons quickly expand.

The hormone gibberellic acid plays an important role in seed germination and early seedling growth. Gibberellic acid induces the synthesis of enzymes required for the metabolism of stored foods, thus providing energy for seedling growth. Gibberellic acid also induces cell division and cell expansion in dark-grown hypocotyls, maintaining their rapid growth through the soil.

APICAL MERISTEMS AND DEVELOPMENT

The early stages of germination simply involve the enlargement of the root, hypocotyl, an d c otyledons that were preform ed in th e embryo. Postembryonic development, however, is focused on the apical meristems. The shoot apical meristem is the source of all the leaves, stems, and their component cells formed during the lifetime of the plant.The meristem itself is composed of a small population of perpetually embryonic (meristematic) cells. These cells grow and divide, giving rise to new cells, but never mature themselves. Thus there is always a source of new cells at the tip of the shoot. The root tip has a similar population of meristematic cells that gives rise to all root tissues. Both of these meristems are characterized by an indeterminate growth pattern: one that is not finite, but, in theory at least, could continue throughout the lifetime of the plant. Apical meristems are involved in several distinct developmental processes.

The meristems are the location of cell proliferation and thus the source of all new cells in the shoot and root systems. The region s below the meristems are the sites of active growth, as new shoot and root tissue rapidly expands. The shoot apical meristem plays a role in organogenesis, the formation of new leaves and axillary buds in a precise spatial pattern. In contrast, the root apical meristem is not involved in organogenesis; lateral roots are initiated by pericycle cells, which are themselves derived from the meristem, usually several centimeters away from the meristem.

The apical meristems also play a role in histogenesis by giving rise to cells that undergo distinct patterns of differentiation to form the specialized tissue types of the shoot and root. While the embryo initially gives rise to the precursors of dermal, ground, and vascular tissues (protoderm, ground meristem, and procambium, respectively), these tissue precursors continue to be formed by the apical meristems and represent the first stages of cell and tissue differentiation.

CELL GROWTH AND CELL DIVISION

Growth is defined as an irreversible increase in mass that is typically associated with an increase in volume. Plant cell growth is associated with meristems and must be carefully regulated in order for organogensis and histogenesis to occur in the appropriate patterns. The plant regulates growth by regulating the extensibility of its cell walls.

A cell that has nonextensible cells walls can take up some water, but eventually the physical pressure of the water inside the cell pressing out on the cell wall (the turgor pressure) prevents the entry of additional water and any further change in volume. In contrast, a cell that has extensible cell walls can take up a substantial volume of water and thus increase in size.

Turgor pressure that would otherwise prevent water entry momentarily decreases because the walls keep stretching.

Typically cell growth occurs in small increments:

- Wall extensibility increases, reducing turgor pressure;
- Reduced turgor pressure allows water to enter the cell, increasing cell volume;
- Wall extensibility decreases, allowing the cell to build up turgor and preventing further water entry
- The cell undergoes a cycle of synthesis of cytoplasmic and wall components, adding to the cell's mass. This cycle of incremental growth is repeated many times until the cell reaches its final size.

The plant hormones auxin and gibberellin are produced in the vicinity of the apical meristems and usually act in concert to induce cell growth. Both hormones regulate wall extensibility, but carry out this function in different ways. Auxin induces the activity of cell membrane H+ adenosine triphosphatase (ATPase) molecules.

Proton (H+ extrusion lowers the pH of the cell wall, thus activating the cell wall enzyme expansin. Expansin cleaves the hydrogen bonds between two cell wall components: The cellulose microfibrils and the hemicellulose molecules that link adjacent cellulose microfibrils. Breakage of these bonds allows these structural wall components to reposition themselves farther apart, increasing wall extensibility.

Gibberellic acid, on the other hand, stimulates the activity of another cell wall enzyme called xyloglucan endotrans-glycosylase (XET). Xyloglucans are a type of hemicellulose that is cleaved by the XET enzyme.

The hormones auxin and gibberellin each promote cell expansion by loosening the bonds between adjacent cell wall molecules. Each hormone acts on a different molecular target. Hemicellulose molecules also allows the cellulose microfibrils to move farther apart, increasing wall extensibility. Cell division and cell growth are often tightly linked.

When the rate of cell division is balanced by cell growth, as in the apical meristems, average cell size does not increase. As the meristem grows away from earlier formed cells, the ratio of growth to division increases, resulting in overall cell enlargement. As the tissues mature further, cell division ceases completely, giving rise to zones of pure cell enlargement where most of the visible growth of the plant occurs.

This relationship between division and growth, coupled with observations of the predictable planes of cell division during histogenesis, indicates that cell division is carefully regulated during plant development. Molecules called cyclin-dependent kinases (CDKs) are key regulators of cell cycling (including cell division) in plants. CDKs are activated by association with a regulatory subunit called a cyclin and by phosphorylation and dephosphorylation events.

The plant hormone cytokinin appears to regulate the cell cycle by interacting with the CDKs. Cytokinins enhance the synthesis of the cyclin subunits that are required for the cell to enter the deoxyribonucleic acid (DNA) synthesis phase of the cell cycle.

Cytokinins also enhance the CDK dephosphorylation step that is required for the cell to progress into mitosis. Both of these processes are inhibited by the hormone abscisic acid; thus a "developmental tug-of-war" occurs between a division-enhancing hormone and a division-suppressing hormone.

The delicate balance between them determines the rate of cell division and this type of interaction is probably typical of the hormonal regulation of many aspects of plant development.

Differentiation

Differentiation is the process whereby cells, tissues, and organs become different from each other and from their precursors. The concept can be applied to organogenesis since, cotyledons, foliage leaves, sepals, and petals may all develop from similar appearing precursors, the leaf primordia.

As these organs mature, they become different from each other in size, shape, and the development of distinctive cell types. For instance, the epidermis tissue of petals is sharply differentiated from that of cotyledons, foliage leaves and sepals that are photosynthetic organs. Correlated with a photosynthetic function, the epidermis of these organs is made up of flat, transparent cells that allow the penetration of light into internal tissues.

Specialized guard cells that allow CO_2 to enter the leaf are also present. In contrast, the epidermal cells of petals contain brightly coloured carotenoid or anthocyanin pigments. These cells also have a papillate shape that imparts a velvetlike sheen to the petal surface. Since, petals carry on minimal photosynthesis, they often lack guard cells. The process of differentiation is best understood on a cellular level.

For instance, guard cells are highly specialized epidermal cells. Early in the development of a leaf, protodermal precursor cells undergo a distinctive pattern of cell divisions. At first the cell divisions are asymmetric, producing one large and one small derivative. The large derivative stops dividing and differentiates as an unspecialized epidermal cell, while the small derivative undergoes another asymmetric division. At an unknown stop signal, the small derivative undergoes a symmetric division, giving rise to two equal sized cells that become the guard cells.

Unlike their plain neighbours, these cells develop a distinctive kidney shape, unevenly thickened cell walls, large, conspicuous chloroplasts, and finally form a pore (the stomatal aperture) between them.

UNIQUENESS OF PLANT DEVELOPMENT

Although plants share many features of development with animals such as apical-basal polarity, regulation of the balance between cell growth and cell division, formation of distinctive patterns of organs, cells and tissues, and differentiation, some aspects of development are unique to plants.

Among these are:

- The formation and maintenance of the perpetually embryonic regions, the apical meristems. The meristems have an indeterminate growth pattern that result in the occurrence of growth, organogenesis, and histogenesis throughout the life of the plant.
- Plant cells have rigid cell walls that prevent cell movement. Thus organogenesis and histogenesis must occur through differential growth and regulation of the planes of cell division. Cell-cell c om munication is im portant in pla nt developmen t, but cell recognition is likely less important than it is in animals since, plant cells keep the same neighbours throughout their life.
- Plant cells are totipotent; that is, able to differentiate as a different cell type if given the appropriate stimulus. Totipotency is likely a reflection of the plant's sedentary lifestyle. Plants can't escape predators and other kinds of damage, but they can readily repair wounds and reconnect vascular strands by differentiating the appropriate cell types.

FERTILIZATION

Fertilization, the process in which gametes—a male's sperm and a female's egg or ovum—fuse together, producing a single cell that develops into an adult organism. Fertilization occurs in both plants and animals that reproduce sexually—that is, when a male and a female are needed to produce an offspring.

This article focuses on animal fertilization. For information on plant fertilization. Fertilization is a precise period in the reproductive process. It begins when the sperm contacts the outer surface of the egg and it ends when the sperm's nucleus fuses with the egg's nucleus. Fertilization is not instantaneous—it may take 30 minutes in sea urchins and up to several hours in mammals. After nuclear fusion, the fertilized egg is called a zygote.

When the zygote divides to a two-cell stage, it is called an embryo. Fertilization is necessary to produce a single cell that contains a full complement of genes. When a cell undergoes meiosis, gametes are formed— a sperm cell or an egg cell. Each gamete contains only half the genetic material of the original cell. During sperm and egg fusion in fertilization, the full amount of genetic material is restored: half contributed by the male parent and half contributed by the female. In humans, there are 46 chromosomes (carriers of genetic material) in each human body cell—except in the sperm and egg, which each have 23 chromosomes.

As soon as fertilization is complete, the zygote that is formed has a complete set of 46 chromosomes containing genetic information from both parents. The fertilization process also activates cell division. Without activation from the sperm, an egg typically remains dormant and soon dies. In general, it is fertilization that sets the egg on an irreversible pathway of cell division and embryo development.

The Fertilization Process

Fertilization is complete when the sperm's nucleus fuses with the egg's nucleus. Researchers have identified several specific steps in this process. The first step is the sperm approaching the egg. In some organisms, sperm just swim randomly towards the egg (or eggs). In others, the eggs secrete a chemical substance that attracts the sperm towards the eggs.

In one species of sea urchin (an aquatic animal often used in fertilization research), the sperm swim towards a small protein molecule in the egg's protective outer layer, or surface coat. In humans there is evidence that sperm are attracted to the fluid surrounding the egg.

The second step of fertilization is the attachment of several sperm to the egg's surface scoat. All animal eggs have surface coats, which are variously named the vitelline envelope (in abalone and frogs) or the zona pellucida (in mammals).

This attachment step may last for just a few seconds or for several minutes. The third step is a complex process in which the sperm penetrate the egg's surface coat. The head, or front end, of the sperm of almost all animals except fish contains an acrosome, a membrane-enclosed compartment. The acrosome releases proteins that dissolve the surface coat of an egg of the same species. In mammals, a molecule of the egg's surface coat triggers the sperm's acrosome to explosively release its contents onto the surface coat, where the proteins dissolve a tiny hole.

A single sperm is then able to make a slitlike channel in the surface coat, through which it swims to reach the egg's cell membrane. In fish eggs that do not have acrosomes, specialized channels, called micropyles, enable a single sperm to swim down through the egg's surface coat to reach the cell membrane.

When more than one sperm enters the egg, the resulting zygote typically develops abnormally. The next step in fertilization—the fusion of sperm and egg cell membranes—is poorly understood. When the membranes fuse, a single sperm and the egg become one cell. This process takes only seconds, and it is directly observable by researchers.

Specific proteins on the surface of the sperm appear to induce this fusion process, but the exact mechanism is not yet known. After fusion of the cell membranes the sperm is motionless. The egg extends cytoplasmic fingers to surround the sperm and pull it into the egg's cytoplasm.

Filaments called microtubules begin to grow from the inner surface of the egg cell's membrane inward towards the cell's centre, resembling spokes of a bicycle wheel growing from the rim inward towards the wheel's hub. As the microtubules grow, the sperm and egg nuclei are pushed towards the egg's centre.

Finally, in a process that is also poorly understood, the egg and sperm nuclear envelopes (outer membranes) fuse, permitting the chromosomes from the egg and sperm to mix within a common space. A zygote is formed, and development of an embryo begins.

Types of Fertilization

Two types of fertilization occur in animals: external and internal. In external fertilization the egg and sperm come together outside of the parents' bodies. Animals such as sea urchins, starfish, clams, mussels, frogs, corals, and many fish reproduce in this way. The gametes are released, or spawned, by the adults into the ocean or a pond. Fertilization takes place in this watery environment, where embryos start to develop. A disadvantage to external fertilization is that the meeting of egg and sperm is somewhat left to chance.

Swift water currents, water temperature changes, predators, and a variety of other interruptions can prevent fertilization from occurring. A number of adaptations help ensure that offspring will successfully be produced. The most important adaptation is the production of litreally millions of sperm and eggs—if even a tiny fraction of these gametes survive to become zygotes, many offspring will still result. Males and females also use behavioural clues, chemical signals, or other stimuli to coordinate spawning so that sperm and eggs appear in the water at the same time and in the same place.

In animals that use external fertilization, there is no parental care for the developing embryos. Instead, the eggs of these animals contain a food supply in the form of a yolk that nourishes the embryos until they hatch and are able to feed on their own. Internal fertilization takes place inside the female's body.

The male typically has a penis or other structure that delivers sperm into the female's reproductive tract. All mammals, reptiles, and birds as well as some invertebrates, including snails, worms, and insects, use internal fertilization. Internal fertilization does not necessarily require that the developing embryo remains inside the female's body.

In honey bees, the queen bee deposits the fertilized eggs into special compartments in the honeycomb. These compartments are supplied with food resources for the young bees to use as they develop. Various adaptations have evolved in the reproductive process of internal-fertilizing organisms.

Because the sperm and egg are always protected inside the male's and female's bodies—and are deliberately placed into close contact during mating—relatively few sperm and eggs are produced. Many animals in this group provide extensive parental care of their young.

In most mammals, including humans, two specialized structures in the female's body further help to protect and nourish the developing embryo. One is the uterus, which is the cushioned chamber where the embryo matures before birth; the other is the placenta, which is a blood-rich organ that supplies nutrients to the embryo and also removes its wastes.

Research Issues

Although reproduction is well studied in many kinds of organisms, fertilization is one of the least understood of all fundamental biological processes. Our knowledge of this fascinating topic has been vastly improved by many recent discoveries. Researchers have discovered how to clone the genes that direct the fertilization process. Yet many important questions still remain.

Scientists are actively trying to determine issues such as how sperm and egg cells recognize that they are from the same species; what molecules sperm use to attach to egg coats; and how signals on the sperm's surface are relayed inside to trigger the acrosome reaction. With continued study, answers to these questions will one day be known.

GAMETE

Gamete, sexual reproductive cell that fuses with another sexual cell in the process of fertilization. The cell resulting from the union of two gametes is called a zygote; the zygote usually undergoes a series of cell divisions until it develops into a complete organism. Gametes, also called germ cells, vary widely in structure. The simplest sexual organisms are isogamous, that is, they produce a single kind of gamete.

The identical gametes unite in pairs to produce zygotes. Although all isogametes are apparently alike in structure, they are thought to be different in physiological constitution, because gametes from the same individual do not successfully unite. The simplest isogametes, those of lower fungi such as molds, are small cells that grow on the ends of body filaments and become detached when mature.

Other lower organisms, such as lower algae and protozoa, have gametes, which are formed by division of the protoplasm of single cells. All higher plants are heterogamous, that is, they produce two kinds of gametes. The female gamete is called the egg; the male gamete is called the sperm.

The organ of gamete production in plants is called a gametangium. All animals and animal-like lower organisms that reproduce sexually, except a few protozoans, a re also heterogamous. The male gametes are called spermatozoa; female gametes, ova or eggs. The gamete-producing organs of animals are called gonads.

The formation of gametes in the gonads of animals is called gametogenesis. By this process the number of chromosomes in the sex cells is reduced in number from diploid to haploid, which is half the number of chromosomes in the normal body cells of the species. The diploid number of human chromosomes. When a human sex cell divides to form two gametes, each ga mete receives only ha lf, or 23, of the norma l com plemen t of chromosomes.

This type of cell division is called meiosis. The normal total of chromosomes is restored in fertilization when two gametes fuse, each contributing half of the chromosomes required by the zygote.

PLANT PROPAGATION

Plant Propagation, growing new plants from seeds or from parts of existing plants. Plant propagation occurs in nature to ensure survival and spread of species. It is also used commercially to produce seeds and plants for agriculture, horticulture, and forestry. Plant propagation includes sexual propagation, which involves the union of sperm and egg to form seeds, and asexual propagation.

Asexual propagation, also known as vegetative propagation, is the growing of new plants from a leaf, stem, or root of a single parent plant. These two forms of plant propagation transmit genetic information between plants of the same species. Genetic engineering transfers genes from one organism to another that may or may not be of the same species to introduce desirable traits into an organism.

These so-called transgenic plants can then be propagated by sexual or asexual propagation.

Sexual Propagation

In nature, sexual propagation begins when water, wind, insects, birds, or small mammals carry pollen randomly between plants. In flowering plants, this transfer of pollen enables the male sex cells, or sperm, of one flower to fertilize the female sex cell, or egg, of a second flower. The egg is located at the base of the flower in a structure called an ovule, found within the ovary. Depending on the species, an ovary contains one, several, or many ovules.

The ovaries of peach and avocado flowers, have one ovule, while those of watermelon and cantaloupe have many. As the fertilized egg (or eggs) within the ovule begins to develop into an embryonic plant, it produces a variety of hormones that stimulate the outer wall of the ovule to harden into a seed coat. Other biochemical changes in the ovule produce a starchy substance that will be used as a food supply.

In this way, the ovule ripens into a seed—a structure containing an embryonic plant and its food supply surrounded by a seed coat. The ovary, which houses the ovule or ovules, is also stimulated by hormones, which cause its tissues to enlarge into a fruit. The fruit contains the ripened ovules, or seeds. In certain nonflowering plants such as pine, spruce, and fir trees, cones, not flowers, hold the ovules.

The young cone, which averages about 2.5 cm (about 1 in) in length, is composed of numerous soft, green, leaf-like scales arranged in a spiral, with two ovules on the upper surface of each scale. Wind transports pollen to the cone scale, where it enters the ovule and fertilizes the egg. The fertilized egg develops into a seed that lies exposed on the cone scale. Seeds, whether fr om f lowerin g o r nonflo wer in g pl an ts , ty pically contai n a r a nd om combination of the genetic material from the sperm and egg of the parents and are therefore genetically unique.

Plants grown from a seed typically have different traits from their parents—they might be weaker, with poor root development or little resistance to disease, or stronger, with good roots and successful resistance to disease. A population with diverse traits is more likely to survive changes in the environment such as emerging diseases, new predatory insects, or climate changes. While environmental fluctuations may kill some plants, those with the necessary adaptations are able to survive, reproduce, and carry on the species.

In this way, traits that help plants survive in different environments get passed down from generation to generation, resulting in the evolution of new species with traits different from those of their ancestors. Sexual propagation also helps plants spread geogra-phically because it produces fruits and seeds that animals such as squirrels and birds consume.

The seeds pass through an animal's digestive system and are deposited at a distance from the parent plants. It is often advantageous for seeds to germinate far from their parents so that parents and offspring are not competing for light, water, and nutrients. Seed companies take advantage of naturally pollinated plants by harvesting seed from the company fields for sale to farmers, home gardeners, and nursery workers.

Seed companies, geneticists, and other researchers also use artificial pollination, or hand pollination, to control the transfer of pollen between plants. In artificial pollination, a paintbrush or other delicate tool is used to transfer pollen between selected plants.

The physical transfer of pollen in this way is called a cross, and the plants that result are also referred to as crosses. Some researchers use artificial pollination to develop new agricultural crops that combine the best traits of both parents. A cross between wheat and rye, produced triticale, a vigourous pasture grass that resists many diseases and insects, and whose seeds are used to make a nutritious flour.

Geneticists also use artificial pollination to study how particular traits are inherited within a species—to determine, if the trait for large seeds is dominant or recessive.

Asexual Propagation

Asexual propagation is the production of new plants from the leaves, stems, or roots of a single parent plant. Asexual propagation, which does not require pollination or fertilization, is a rapid method of propagation.

It ensures that all of the parent's genetic material survives even if the parent dies, and it creates offspring, known as clones, with the same traits as the parent plant. Asexual propagation is advantageous when plants are well adapted to a particular environment. Several methods of asexual propagation occur in nature. They have been adapted for commercial use for rapid propagation and to obtain plants that are hard to grow from seeds.

Cuttings and Layering

For many plant species, a leaf, section of stem, or piece of root cut from a plant and lightly covered in soil, peat moss, or another growth medium develops a new, independent plant by generating the missing parts. Stimulated by hormones called auxins, a partially buried leaf or piece of stem, develops roots on the buried portion, and a piece of root forms stems and leaves above the soil.

Cuttings and layering are widely used for perennial plants, plants that grow back from the same roots year after year. Commercially, cuttings are the most important source for perennials, such as new fruit trees; conifers, including pine and spruce; a variety of shrubs, roses and honeysuckle, and many florist blooms.

In layering, a new plant develops from a stem that is still attached to the parent plant. In nature, the stem simply arches over and spreads out on the ground. The parts of the stem that are in secure contact with the soil develop the roots, stems, and leaves of a new plant. The flexible stems of trailing blackberries, black raspberries, and several other species spread rapidly in this way.

Growers typically layer species that propagate naturally in this way. Layering is also used for plants such as filberts and Muscadine grapes that cannot be propagated easily by other methods. If plants do not layer naturally, a grower can induce them to layer by pinning the stems to the ground. Growers also layer plants by cutting them to the ground during the winter and covering the new spring shoots with soil, a technique called mound layering. This causes new roots to form along the buried lower portion of the new shoots, and new stems and leaves then grow up through the mound of soil. In a few weeks the new plants are cut away from the parent and planted in a garden or nursery.

This technique enables a grower to produce many plants from a single parent that has desired traits. Apple trees, currants, and gooseberries are commonly propagated in this way. In air layering, often used with house plants, stems are partially cut and peat moss or sphagnum moss is wrapped around the wound, which is then covered in plastic wrap. A single plant develops from the place where the stem was cut.

Grafting

In grafting, a freshly cut section of stem with buds, called a scion, is joined to another plant called the stock. The upper stem of the stock is severed and the scion is joined to the lower stem. The scion is securely attached to the stock, and the tissues of the two plants grow into each other, forming a single plant.

The scion produces the stems, leaves, and flowers on the new plant and the stock provides the root system. Grafting combines desirable qualities from one species, such as disease resistance or the ability to grow in waterlogged soils, with desirable qualities of another, such as the ability to produce high quality fruit.

Grafting is often used to make fruit trees more vigo urous a n d productive. Bud grafting is a form of grafting in which a single bud cut from a stem is grafted onto the stock. It can be carried out more rapidly than other forms of grafting and is used widely in the nursery industry to propagate hundreds or thousands of plants in a relatively short amount of time.

In nature, roots of oak trees of the same species commonly graft together, hastening the spread of diseases such as oak wilt, a fungal disease that kills a variety of oak trees.

Agamospermy

In agamospermy, also known as apomixis, a seed develops directly from tissues of the ovule rather than from a fertilized egg. Depending on the species, a fruit may or may not be produced. The plants that develop by agamospermy are clones of the mother plant. Agamospermy occurs in nature in species such as dandelions and blackberries, enabling them to spread rapidly since, they can bypass pollination and fertilization.

Tissue Culture

Also called micropropagation, tissue culture is the production of plants under sterile labouratory conditions. A variety of tissue culture techniques are used to propagate plants. In one method, growers remove a tiny piece of leaf or stem from a plant and place it in a sterile test tube on a gel-like medium enriched with hormones and nutrients. A yellow-brown mass of cells called callus develops from the piece of plant.

Small chunks of the callus are separated, and each piece is placed in a petridish with a hormone and nutrient mix that stimulates the development of the call us pieces into plants. The young plants are removed from the petridish and placed in pots with soil, or into the ground, where they grow to maturity. Tissue culture enables researchers and growers to rapidly generate numerous clones year-round in greenhouses.

In nature, strawberry plants typically produce their fruits in summer. Commercially grown strawberries, however, are propagated throughout the year by tissue culture, providing consumers with a steady supply of strawberries for every season. Tissue culture is also used to produce plants free of viruses, fungi, and bacteria, and to propagate species such as Douglas fir and rhododendron, which are difficult to grow commercially from cuttings, layering, or grafting.

Propagation from Stems and Roots

Some plants produce special underground stems such as tubers, bulbs, and corms that enable them to reproduce asexually. Like all stems, these structures have buds, or nodes, from which new stems branch. Tubers are swollen, fleshy stems with several buds called eyes that produce new plants; an example of a tuber is the potato. Bulbs, such as those found in onions, lilies, hyacinths, and tulips, are short, wide, teardrop-shaped underground stems surrounded by scaly leaves. Corms, such as crocuses and gladioli, are similar, but lack the scaly leaves.

Both bulbs and corms make clumps of new bulbs or corms, called offsets, which can be divided and buried in the soil to generate new plants. Irises and ferns produce rhizomes, fleshy stems that grow horizontally beneath the soil, with new plants developing from the tip of the rhizome and from each node on the stem.

Stolons, specialized stems found in strawberries and many lawn grasses, are similar to rhizomes but are usually thinner and grow on top of the soil. They also produce new plants at the tip and from the nodes.

REPRODUCTION IN PLANTS

Plant reproduction is the process by which plants generate new individuals, or offspring. Reproduction is either sexual or asexual. Sexual reproduction is the formation of offspring by the fusion of gametes. Asexual reproduction is the formation of offspring without the fusion of gametes. Sexual reproduction results in offspring genetically different from the parents.

Asexual offspring are genetically identical except for mutation. In higher plants, offspring are packaged in a protective seed, which can be long lived and can disperse the offspring some distance from the parents. In flowering plants (angiosperms), the seed itself is contained inside a fruit, which may protect the developing seeds and aid in their dispersal.

Sexual Reproduction in Angiosperms: Ovule Formation

All plants have a life cycle that consists of two distinct forms that differ in size and the number of chromosomes per cell. In flowering plants, the large, familiar form that consists of roots, shoots, leaves, and reproductive structures (flowers and fruit) is diploid and is called the sporophyte. The sporophyte produces haploid microscopic gametophytes that are dependent on tissues produced by the flower. The reproductive cycle of a flowering plant is the regular, usually seasonal, cycling back and forth from sporophyte to gametophyte. The flower produces two kinds of gametophytes, male and female.

The female gametophyte arises from a cell within the ovule, a small structure within the ovary of the flower. The ovary is a larger structure within the flower that contains and protects usually many ovules. Flowering plants are unique in that their ovules are entirely enclosed in the ovary. The ovary itself is part of a larger structure called the carpel, which consists of the stigma, style, and ovary. Each ovule is attached to ovary tissue by a stalk called the funicle. The point of attachment of the funicle to the ovary is called the placenta. As the flower develops from a bud, a cell within an ovule called the archespore enlarges to form an embryo-sac mother cell (EMC). The EMC divides by meiosis to produce four megaspores. In this process the number of chromosomes is reduced from two sets in the EMC to one set in the megaspores, making the megaspores haploid. Three of the four megaspores degenerate a nd disappear, while the fourth divides mitotically three times to produce eight haploid cells. These cells together constitute the female gametophyte, called the embryo sac. The eight embryo sac cells differentiate into two synergids, three antipodal cells, two fused endosperm nuclei, and an egg cell. The mature embryo sac is situated at the outer opening (micropyle) of the ovule, ready to receive the sperm cells delivered by the male gametophyte.

Pollen

The male gametophyte is the mature pollen grain. Pollen is produced in the anthers, which are attached at the distal end of filaments. The filament and anther together constitute the stamen, the male sex organ. Flowers usually produce many stamens just inside of the petals. As the flower matures, cells in the anther divide mitotically to produce pollen mother cells (PMC).

The PMCs divide by meiosis to produce haploid microspores in groups of four called tetrads. The microspores are housed within a single layer of cells called the tapetum, which provides nutrition to the developing pollen grains.

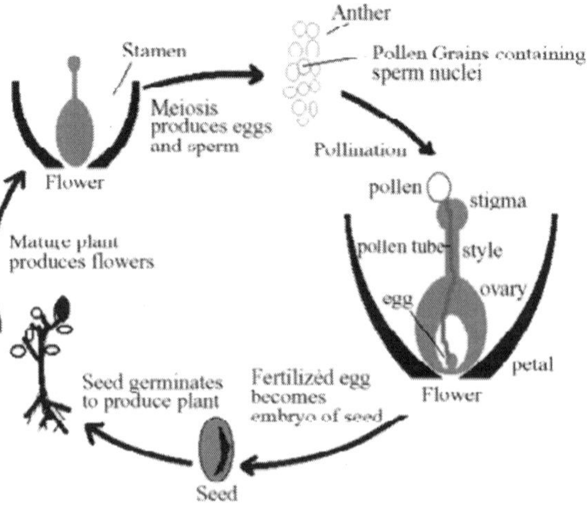

Fig.Pollen

Each microspore develops a hard, opaque outer layer called the exine, which is constructed from a lipoprotein called sporopollenin. The exine has characteristic pores, ridges, or projections that can often be used to identify a species, even in fossil pollen. The microspore divides mitotically once or twice to produce two or three haploid nuclei inside the mature pollen grain.

Two of the nuclei function as sperm nuclei that can eventually fuse with the egg and endosperm nuclei of the embryo sac, producing an embryo and endosperm, respectively. For sexual fusion to take place, however, the pollen grain must be transported to the stigma, which is a receptive platform on the top of the style, an elongated extension on top of the carpel(s).

Here the moist surface or chemicals cause the pollen grain to germinate. Germination is the growth of a tube from the surface of a pollen grain. The tube is a sheath of pectin, inside of which is a solution of water, solutes, and the two or three nuclei, which lack any cell walls. Proper growth of the pollen tube requires an aqueous solution of appropriate solute concentration, as well as nutrients such as boron, which may aid in its synthesis of pectin. At the apex of the tube are active ribosomes and endoplasmic reticulum (types of cell organelles) involved in protein synthesis.

Pectinase and a glucanase (both enzymes that break down carbohydrates) probably maintain flexibility of the growing tube and aid in penetration. The pollen tube apex also releases ribonucleic acid (RNA) and ribosomes into the tissues of the style.

The tube grows to eventually reach the ovary, where it may travel along intercellular spaces until it reaches a placenta. Through chemical recognition, the pollen tube changes its direction of growth and penetrates through the placenta to the ovule. Here the tube reaches the embryo sac lying close to the micropyle, and sexual fertilization takes place.

Double Fertilization

Fertilization in flowering plants is unique among all known organisms, in that not one but two cells are fertilized, in a process called double fertilization. One sperm nucleus in the pollen tube fuses with the egg cell in the embryo sac, and the other sperm nucleus fuses with the diploid endosperm nucleus.

The fertilized egg cell is a zygote that develops into the diploid embryo of the sporophyte. The fertilized endosperm nucleus develops into the triploid endosperm, a nutritive tissue that sustains the embryo and seedling. The only other known plant group exhibiting double fertilization is the Gnetales in the genus Ephedra, a nonflowering seed plant.

However, in this case the second fertilization product degenerates and does not develop into endosperm. Double fertilization begins when the pollen tube grows into one of the two synergid cells in the embryo sac, possibly as a result of chemical attraction to calcium. After penetrating the synergid, the apex of the pollen tube breaks open, releasing the two sperm nuclei and other contents into the synergid. As the synergid degenerates, it envelops the egg and endosperm cells, holding the two sperm nuclei close and the other expelled contents of the pollen tube. The egg cell then opens and engulfs the sperm cell, whose membrane breaks apart and allows the nucleus to move near the egg nucleus. The nuclear envelopes then disintegrate, and the two nuclei combine to form the single diploid nucleus of the zygote.

The other sperm cell fuses with the two endosperm nuclei, forming a single triploid cell, the primary endosperm cell, which divides mitotically into the endosperm tissue. Double fertilization and the production of endosperm may have contributed to the great ecological success of flowering plants by accelerating the growth of seedlings and improving survival at this vulnerable stage.

Faster seedling development may have given flowering plants the upper hand in competition with gymnosperm seedlings in some habitats, leading to the abundance of flowering plants in most temperate and tropical regions. Gymnosperms nevertheless are still dominant at higher elevations and latitudes, and at low elevations in the Pacific Northwest coniferous forests, such as the coastal redwoods. The reasons for these patterns are still controversial.

THE SEED

The seed is the mature, fertilized ovule. After fertilization, the haploid cells of the embryo sac disintegrate. The maternally derived diploid cells of the ovule develop into the hard, water-resistant outer covering of the seed, called the testa, or seed coat. The diploid zygote develops into the embryo, and the triploid endosperm cells multiply and provide nutrition.

The testa usually shows a scar called the hilum where the ovule was originally attached to the funicle. In some seeds a ridge along the testa called the raphe shows where the funicle originally was pressed against the ovule. The micropyle of the ovule usually survives as a small pore in the seed coat that allows passage of water during germination of the seed.

In some species, the funicle develops into a larger structure on the seed called an aril, which is often brightly coloured, juicy, and contains sugars that are consumed by animals that may also disperse the seed (as in nutmeg, arrowroot, oxalis, and castor bean). This is distinct from the fruit, which forms from the ovary itself. The embryo consists of the cotyledon(s), epicotyl, and hypocotyl. The cotyledons resemble small leaves, and are usually the first photosynthetic organs of the plant.

The portion of the embryo above the cotyledons is the epicotyl, and the portion below is the hypocotyl. The epicotyl is an apical meristem that produces the shoot of the growing plant and the first true leaves after germination. The hypocotyl develops into the root. Often the tip of the hypocotyl, the radicle, is the first indication of germination as it breaks out of the seed. Flowering plants are classified as monocotyledons or dicotyledons (most are now called eudicots) based on the number of cotyledons produced in the embryo. Common monocotyledons include grasses, sedges, lilies, irises, and orchids; common dicotyledons include sunflowers, roses, legumes, snapdragons, and all nonconiferous trees. The endosperm may be consumed by the embryo, as in man y le gumes, which use the cotyledons as a food source during germination.

In other species the endosperm persists until germination, when it is used as a food reserve. In grains such as corn and wheat, the outer layer of the endosperm consists of thick-walled cells called aleurone, which are high in protein.

THE FRUIT

The fruit of a flowering plant is the mature ovary. As seeds mature, the surrounding ovary wall forms a protective structure that may aid in dispersal. The surrounding ovary tissue is called the pericarp and consists of three layers. From the outside to inside, these layers are the exocarp, mesocarp, and endocarp. The exocarp is usually tough and skinlike.

The mesocarp is often thick, succulent, and sweet. The endocarp, which surrounds the seeds, may be hard and stony, as in most species with fleshy fruit, such as apricots. A fruit is termed simple if it is produced by a single ripened ovary in a single flower (apples, oranges, apricots). An aggregated fruit is a cluster of mature ovaries produced by a single flower (blackberries, raspberries, strawberries).

A multiple fruit is a cluster of many ripened ovaries on separate flowers growing together in the same inflorescence (pineapple, mulberry, fig). A simple fruit may be fleshy or dry. A fleshy simple fruit is classified as a berry (grape, toma to, pa pa ya) , pe po (c ucum ber, w a term e l on, pum p ki n), hesperidium (orange), drupe (apricot), or pome (apple). Dry simple fruits have a dry pericarp at maturity. They may or may not dehisce, or split, along a seam to release the seeds.

A dehiscent dry fruit is classified as legume or pod (pea, bean), silique or silicle (mustard), capsule (poppy, lily), or follicle (milkweed, larkspur, columbine). An indehiscent dry fruit that does not split to release seeds is classified as an achene (sunflower, buttercup, sycamore), grain or caryopsis (grasses such as corn, wheat, rice, barley), schizocarp (carrot, celery, fennel), winged samara (maple, ash, elm), nut (acorn, chestnut, hazelnut), or utricle (duckweed family).

Some fruiting bodies contain non-ovary tissue and are sometimes called pseudocarps. The sweet flesh of apples and pears, is composed not of the pericarp but the receptacle, or upper portion, of the flowering shoot to which petals and other floral organs are attached. Fruiting bodies of all kinds function to protect and disperse the seeds they contain.

Protection can be physical (hard coverings) or chemical (repellents of seed predators). Sweet, fleshy fruits are attractive food for birds and mammals that consume seeds along with the fruit and pass the seeds intact in their fecal matter, which can act as a fertilizer. Dry fruits are usually adapted for wind dispersal of seeds, with the assistance of winglike structures or a fluffy pappus that provides buoyancy.

The diversity of fruiting bodies reflects in part the diversity of dispersal agents in the environment, which select for different fruit size, shape, and chemistry.

POLLINATION AND POLLINATORS

Pollination is the movement of pollen from the stamens to the stigma, where germination and growth of the pollen tube occur. Most (approximately 96 per cent) of all flowering plant species are hermaphroditic (possess both sexual functions within a plant, usually within every flower), and thus an individual can be pollinated by its own pollen or by pollen from another individual.

Seed produced through self-pollination ("selfed" seed) is often inferior in growth, survival, and fecundity to seed produced through outcross pollination ("outcrossed" seed). As a result, in most species there is strong natural selection to maximize the proportion of outcrossed seed (the "outcrossing rate"). Flowering plants are unusual among seed plants in their superlative exploitation of animals (primarily insects) as agents of outcross pollination.

The outcross pollination efficiency of insects, birds, and mammals (primarily bats) may have contributed to both the abundance and diversity of flowering plants. Abundance may have increased because of less wastage of energy and resources on unsuccessful pollen and ovules. Diversity may have increased for two reasons. First, insects undoubtedly have selected for a wide variety of floral forms that provide different rewards (pollen and nectar) and are attractive in appearance (colour juxtaposition, size, shape) and scent (sweet, skunky) in different ways to different pollinators.

Second, faithfulness of pollinators to particular familiar flowers may have reduced hybridization and speeded evolutionary divergence and the production of new species. Although flowering plants first appeared after most of the major groups of insects had already evolved, flowering plants probably caused the evolution of many new species within these groups.

Some new insect groups, such as bees and butterflies, originated after flowering plants, their members developing mouthpart structures and behaviour specialized for pollination. In extreme cases, a plant is completely dependent on one insect species for pollination, and the insect is completely dependent on one plant species for food.

Such tight interdependency occurs rarely but is well documented in yuccas/yucca moths, senita cacti/senita moths, and fig trees/some fig wasps. In all three insects, females lay eggs in the flowers, and their young hatch later to feed on the mature fruit and its contents. Females ensure that the fruit develops by gathering pollen from another plant and transporting it to the stigma of the flower holding their eggs. Plants benefit greatly in outcrossed seed produced, at the small cost of some consumed fruit and seeds, and the insects benefit greatly from the food supply for developing larvae at the small cost of transporting pollen the short distances between plants.

Pollinating agents, whether biotic or abiotic, have exerted strong selection on all aspects of the flower, resulting in the evolution of tremendous floral diversity. This diversity has been distilled into a small number of characteristic pollination syndromes. Pollination by beetles selects usually for white colour, a strong fruity scent, and a shallow, bowl-shaped flower.

Bees select for yellow or blue/purple colourings, a landing platform with colour patterns that guide the bee to nectar (often reflecting in the ultraviolet range of the spectrum), bilateral symmetry, and a sweet scent. Butterflies select for many colours other than yellow, a corolla (petal) tube with nectar at the base, and the absence of any scent. Moths in contrast select for nocturnally opening flowers with a strong scent and drab or white colour, and also a tube with nectar at the base.

Bats select also for nocturnally opening flowers, but with a strong musky scent and copious nectar, positioned well outside the foliage for easy access, and drab or white colour. Hummingbirds select for red or orange flowers with no scent, copious nectar production, and a corolla tube with nectar at the base.

Other pollinating birds that do not hover while feeding select for strong perches and flowers capable of containing copious nectar (tubes, funnels, cup shapes). Wind as a pollinating agent selects for lack of colour, scent, and nectar; small corolla; a large stigmatic surface area (usually feathery); abundantly produced, buoyant pollen; and usually erect styles and limp, hanging stamens. In addition there is great floral diversity within any of these syndromes, arising from the diverse evolutionary histories of the member plant species.

Selfing and Outcrossing

Most flowering plant species reproduce primarily by outcrossing, including the great majority of trees, shrubs, and perennial herbs. Adaptations that prevent self-fertilization include self-incompatibility (genetic recognition and blocking of self-pollen) and dioecy (separate male and female individuals).

Adaptations that reduce the chances of self-pollination in hermaphrodites include separation of the anthers and stigma in space (herkogamy) or time (dichogamy). In many species, both self-incompatibility and spatiotemporal separation of the sex organs occur. The ability to produce seeds by selfing, however, is advantageous in situations where outcrossing pollination is difficult or impossible. These include harsh environments where pollinators are rare or unpredictable, and regularly disturbed ground where survivors often end up isolated from each other. Selfing is also cheaper than outcrossing, because selfers can become pollinated without assistance from animals and therefore need not produce large, attractive flowers with abundant nectar and pollen rewards. Most primarily selfing species are small annuals in variable or disturbed habitats, with small, drab flowers. Most desert annuals and roadside weeds are selfers. The evolutionary transition from outcrossing to near-complete selfing has occurred many times in flowering plants. Outcrossing and selfing species differ in their evolutionary potential. Outcrossers are generally more genetically diverse and produce lineages that persist over long periods of evolutionary time, during which many new species are formed. Selfers, however, are less genetically diverse and tend to accumulate harmful mutations. They typically go extinct before they have an opportunity to evolve new species.

Asexual Reproduction

The ability to produce new individuals asexually is common in plants. Under appropriate experimental conditions, nearly every cell of a flowering plant is capable of regenerating the entire plant. In nature, new plants may be regenerated from leaves, stems, or roots that receive an appropriate stimulus and become separated from the parent plant. In most cases, these new plants arise from undifferentiated parenchyma cells, which develop into buds that produce roots and shoots before or after separating from the parent. New plants can be produced from aboveground or belowground horizontal runners (stolons of strawberries, rhizomes of many grasses), tubers (potato, Jerusalem artichoke, dahlia), bulbs (onion, garlic), corms (crocus, gladiola), bulbils on the shoot (lily, many grasses), parenchyma cells in the leaves (Kalanchoe, African violet, jade plant) and inflorescence (arrowhead).

Vegetative propagation is an economically important means of replicating valuable agricultural plants, through cuttings, layering, and grafting. Vegetative reproduction is especially common in aquatic vascular plants (example, surfgrass and eelgrass), from which fragments can break off, disperse in the current, and develop into new whole plants. A minority of flowering plants can produce seeds without the fusion of egg and sperm (known as parthenocarpy or agamospermy). This occurs when meiosis in the ovule is interrupted, and a diploid egg cell is produced, which functions as a zygote without fertilization. Familiar examples include citrus, dandelion, hawkweed, buttercup, blackberry/raspberry, and sorbus. Agamospermous species are more common at high elevations and at high latitudes, and nearly all have experienced a doubling of their chromosome number (tetraploidy) in their recent evolutionary history. These species experience evolutionary advantages and disadvantages similar to those of selfers.

Evolutionary Significance of Plant Reproduction Strategies

The attractive, colourful, and unique features of the most abundant and diverse group of land plants—the flowering plants—are believed to have evolved primarily to ma ximize the efficien cy an d speed of outcross reproduction. Each major burst of angiosperm evolution was a coevolutionary episode with associated animals, primarily insects, which were exploited to disperse pollen and seeds in ever more efficient and diverse ways. The first major burst of flowering plant evolution was the appearance of the closed carpel together with showy flowers that were radially symmetrical.

The closed carpel prevented self-fertilization through recognition and blocking of self pollen within the specialized conducting tissue of the style. Insects attracted to the showy flowers carried pollen between plants less wastefully than wind, and the radial symmetry accommodated insects of many sizes and shapes. The second major burst was the appearance of bilaterally symmetrical flowers, which happened independently in many groups of plants at the same time that bees evolved.

Bilateral symmetry forced bees to enter and exit flowers more precisely, promoting even more efficient outcross pollen transfer. The third major burst of flowering plant evolution was the appearance of nutritious, fleshy fruits and seeds, coincident with a diversification of birds and rodents.

The exploitation of vertebrates for fruit and seed dispersal resulted in less haphazard transport of offspring to neighbouring populations of the same species (also visited as a food source), thereby reducing the chances that progeny inbreed with their siblings and parents and providing more assurance than wind currents that they find good habitat and unrelated mating partners of the same species.

❐

Index

❑❑❑